Analysis of Data

Analysis of Data

SECOND EDITION

J. M. Haile

MACATEA PRODUCTIONS • CENTRAL, SOUTH CAROLINA

Published by Macatea Productions, Central, South Carolina, USA.
http://www.macatea.com/
contact@macatea.com

First edition published 2003
Second edition published 2009

ISBN: 978-0-9728602-3-9
Library of Congress Control Number: 2009906302

Publisher's Cataloging-in-Publication Data:

Haile, J. M.
    Analysis of data / J. M. Haile
    p. cm.
    Includes index.
    ISBN 978-0-9728602-3-9

    1. Mathematical statistics–Methodology. I. Title

QA276.H23 2009   001.4'22
                      QBI03-200150

12 11 10 09 08 07

Preface

To ENGINEER IS TO EXERCISE JUDGMENT—to make decisions—and it is commonly held that good decisions require good data; but this is a fallacy. Good decisions depend on good analyses, and good analyses may be performed on data that are bad or good. This implies that an analysis is more important than the data being analyzed: analysis adds value to data.

We collect data to test models, to confirm hypotheses, to strengthen designs, to connect effect to cause, to meet regulatory and safety demands, to advance knowledge. Students, faculty, technicians, and practicing engineers may spend weeks, months, years designing experiments, building apparatus, calibrating equipment, recording data. But then too often, the fruits of those efforts are condensed into superficial reports based on perfunctory analyses. In many cases the analysis is little more than a plot, a least-squares fit, and an assertion that the data behave as expected or that the data agree with the prevailing theory. Given the dedicated efforts made in obtaining the data, it seems to me that a comparable degree of dedication ought to be invested in squeezing the data for as much meaning as possible. These observations motivate this book.

The intention is to help you start an analysis and to remind you that getting the data is only one part of the adventure. Of course, a huge literature already exists on analysis—statistical, exploratory, confirmatory, and otherwise. But it might be helpful to have a little introduction that reminds us where to start and what fundamental analyses should be attempted. Those are the topics covered here. They divide into five groups, which form the chapters that follow.

When first confronting data, we should assess the quality of the measurements (Chapter 2), for little meaning can be attached to data of unknown quality. Our treatment follows the internationally accepted guidelines for expressing uncertainties.

With the quality established, we may begin to quantify the relations between a dependent variable y and an independent variable x (Chapter 3). Since this is only an introduction, we limit our attention to functions of a single variable. Since an engineer's thinking is dominated by straight lines, we focus on finding transformations that linearize the relations. Of course, not all relations are linear, but this is how we start:

we try straight lines first and abandon them reluctantly. In some situations, we don't even need a straight line, we only need to establish whether x and y are in fact correlated (Chapter 4).

Both in the design of an experiment and in the analysis of its data, we benefit from knowing how y responds to changes in the independent variables (Chapter 5). If y is insensitive to some variables, then those variables need be controlled only loosely or measured relatively imprecisely. The remaining variables will require more attention in the lab and in the analysis.

Finally, we offer examples of activities that can carry us beyond a basic analysis toward deeper meaning (Chapter 6). Again, the activities discussed are intended to help start a deeper analysis; they are far from comprehensive.

The difficulty of performing an analysis correlates only weakly with the difficulty of performing the experiment. A difficult experiment may yield to simple analysis; a simple experiment may require difficult tedious analysis. However, analysis of data is closely related to the design of the experiment; in fact, I believe there is a kind of conservation law that usually applies. You can devote considerable effort in the design stage, so that the analysis is straightforward, or you can skimp on the design, leaving much effort to be devoted to teasing meaning from the data. In other words, if you want a straightforward analysis, then you must plan for it before you enter the lab. Perhaps this little book can guide your planning, for planning is where informative analyses usually start.

Preface to the Second Edition

The principal changes in this second edition are the additions of § 6.4.4, Appendix B, and new exercises at the end of each chapter. I caution students and instructors that some of the new exercises are open-ended in the sense that I have not necessarily provided, in some problem statements, all the information needed to find a solution. In those cases, students and instructors should appeal to their own experience, to common knowledge, and to standard reference works to close the problem statement: think outside the book.

Contents

List of Exhibits

Analysis of Data

If one wishes to learn but little,
it is true that a very little thought is enough,
but if we want to learn more,
we must think more.

John W. Tukey, *Exploratory Data Analysis*, 1977, p. 141.

Exhibit A

Another difficulty the nineteenth-century cook had to face was the imprecision of her cookbooks. A criticism often made of them today is that they failed to specify how much of each ingredient should go into a dish, either by weight or measure. Usually they did so because it would have been useless. Accurately standardized measuring cups and spoons were lacking and few kitchens were equipped with scales and weights. ... Most cookbooks did not bother with such details.

from W. Root and R. de Rochemont,
Eating in America. A History,
Morrow, New York, 1976, p. 138.

Table of Weights and Measures
for those persons who have neither scales nor weights

1 lb. of wheat-flour = 1 quart

1 lb. 2 oz. of Indian meal = 1 quart

1 lb. of butter when soft = 1 quart

1 lb. 1 oz. of powdered loaf-sugar = 1 quart

10 eggs = 1 pound

Liquids

8 large tablespoonfuls = 1 gill

1 common-sized tumbler = 1/2 pint

1 common-sized wineglass = 1/2 gill

from Mrs. J. Chadwick, *Home Cookery:*
A Collection of Tried Recipes, both Foreign and Domestic,
Crosby, Nichols, and Co., Boston, 1853.
Reprinted by Arno Press, New York, 1973.

1

The Underpinnings of Analysis

A NALYSIS IS THE ACTIVITY BY WHICH we attach meaning to measured data. To find meaning, analysts must be able to distinguish measurement from experiment (§ 1.1); they must understand the theoretical context that provides the basis for the experiment (§ 1.2); they must know how to resolve the distinctions between reality and the models that are used to represent the experimental situation (§ 1.3); and they must be aware of alternatives to measurement and experiment (§ 1.4).

1.1   Measurement Is Not Experiment

We distinguish experiment from observation and measurement:

OBSERVATION is the application of one or more human senses (seeing, hearing, touching, smelling; tasting is generally to be avoided) to identify qualitative relations among selected objects, events, and behaviors. Typically, an observation leads us to propose a cause-and-effect relation. For example, we observe that a glass of water thrown onto a sleeping cat provokes a dramatic reaction.

MEASUREMENT refers to a physical process by which we assign numbers to quantities. We measure when building a house, when following a cooking recipe, when monitoring personal health.

EXPERIMENT combines observation with measurement to quantify relations. In an experiment we observe relations, we measure to quantify relations, and we analyze to add meaning to relations.

The goal of measurement is to obtain numbers; the goal of experiment is to identify, explore, confirm, and quantify relations. As suggested in Figure 1.1, it is analysis that carries us from numbers into the meaning embedded in relations. Without analysis, there is no experiment, only measurement.

This means that an experiment is performed not merely to quantify, but to obtain quantitative relations that allow us to explain observations

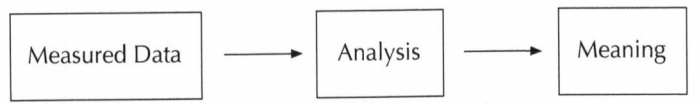

FIGURE 1.1 Analysis is the part of experiment that carries us from numbers for quantities to the meaning embedded in relations among those quantities.

and reason from them [1]. In general, we expect that the more precise and accurate the quantification, the more discriminating and closely reasoned the explanations can be. It would seem that we seek definitive statements of this form:

If A = 100.0, then as a consequence, B will occur.

But often, especially in engineering, what we really need to know is how much uncertainty can be tolerated. We often perform experiments to be able to make statements like this [1]:

If A is close to 100, then B will probably occur.

In these situations, the purpose of an experiment is to quantify "close to" and "probably." Uncertainties not only allow us to judge the quality of measurements, but they may allow us to assess sensitivity and tolerance.

1.2   Theoretical Context

A theoretical context is crucial to both the planning of an experiment and the interpretation of its results. During planning, a theoretical context suggests which quantities are likely to be important: which should be measured, which can be controlled, which can be ignored. But these are only suggestions; if the theoretical context is complete and valid, then we don't need to perform the experiment.

A theoretical context suggests the ranges of values that are likely to be encountered for important quantities. These guide the sizing of equipment and choosing of instrumentation. Further, a theoretical context may help us estimate magnitudes for quantities that affect experimental design and protocols, quantities such as sensitivities, uncertainties, and signal-to-noise ratios. During analysis, a theoretical context might suggest functional forms for relations among variables and it may guide a search for new relations, thereby refining the theory and adding to our understanding.

A complete understanding of the theoretical context requires that we grasp differences in levels of theoretical detail that apply to the experiment [2]. This means we must be able to identify scales over which phenomena should appear, and we must be able to estimate magnitudes of phenomena that might compete with or distract us from the behavior that is to be studied. As the required accuracy of measurements increases, such phenomena tend to become more subtle, requiring more sophisticated theories to deal with them.

1.3 Models

A model is an abstract representation of a relation (or set of relations) that attempts to preserve important features while suppressing unimportant details. Often, a model can be represented by a mathematical equation, such as the ideal-gas law,

$$PV = nRT \tag{1.1}$$

but a model may also be represented by a drawing or a plot. Regardless of the form a model takes, a model should be as simple as possible, yet still capture those features that are important to the situation in which it is being used. Models enter an experiment at several stages:

(a) The theoretical context is usually posed as one or more models. Examples include theoretical stages in mass-transfer operations, incompressible fluids, and Raoult's law. Such models may be crude or sophisticated, but in either case, one purpose of the experiment might be to test or refine such models.

(b) The apparatus is often represented by a model. That is, the experimentalist deals with two pieces of equipment [1]: the real apparatus in the lab and an idealized model that is carried in the head.

A schematic diagram represents the model, not the real apparatus. In the real equipment, such things as nuts, bolts, flanges, packing, and structural supports are needed to hold the equipment together. Usually, such hardware is irrelevant to the theoretical context and is abstracted away in the model apparatus. However, in assessing uncertainties in data, attention must be paid to quantifying the extent to which the real apparatus deviates from its model. This may require us to assess instrument bias, calibration errors, dead volumes in valves and piping, leaking fittings, line-voltage fluctuations, impurities, plus the idiosyncrasies of computers and personnel that monitor the experiment and record data.

(c) Experimental results are commonly represented by models; first consider the extremes. At one extreme we find empirical models, which are mere fits of data to convenient mathematical forms; such fits are legitimately used to communicate and interpolate results. At the other extreme are theoretical models, derived from some fundamental theory. These establish relations among variables, relations that are independent of the experiment. Such models can be used to interpolate, extrapolate, and interpret the data.

Between these extremes we find a spectrum of models that combine some empiricism with some theory. These might be called semiempirical or semitheoretical. One common example occurs when a theory is used to provide a functional form for relations among variables, then experimental data are used to assign (or fit) numerical values to parameters in those functions.

1.4   Other Ways to Get Numbers

Since measurement and experiment are costly and time-consuming, it is worthwhile to consider alternatives. Even if, in a particular situation, these alternatives prove unsatisfactory, they can still help us improve plans for the subsequent experimental design and analysis.

1.4.1   Alternatives to Measurement

Here are common ways to get numbers, without measuring them:

EXTRAPOLATE. Extrapolations of theoretical models can be useful, provided you can maintain the conditions under which the theory applies. Extrapolations of empirical models can be misleading and are to be avoided.

COMPUTE BOUNDS. Bounds are often provided by extreme models. In some situations a knowledge of bounds is sufficient to resolve your problem. In other situations the bounds may be so tight that they effectively provide a reliable numerical value. Even loose bounds can be instructive, for at least they may help us identify ways *not* to think about the problem.

GUESS. In this digital age, to guess is quite unfashionable, but the only bad guess is the one we make uncritically [3]. That is, a blind guess is foolish, but a reasoned guess, however wrong, can lead us to a better guess; and so, a guess can serve to focus our thinking.

We may, of course, go further by combining two or even all three of these alternatives; for example, we might compute bounds and then make a reasoned guess within those bounds.

1.4.2 Alternatives to Experiment

Instead of performing experiments to establish relations, we might apply one (or more) of the following:

FORMAL THEORIES might enable us to derive the relations we seek.

ANALOGIES might be drawn from other experimental situations.

SEMITHEORETICAL MODELS, such as corresponding states and scaling laws, might let us generalize or extend existing data and relations to our situation.

COMPUTER SIMULATIONS might be performed. The pitfall to be avoided is confusing simulation results with real data. Simulations are performed on models, and the relations they provide are among model variables. We then have the separate tasks either of validating that the model faithfully represents reality or of documenting the distinctions between the model and reality.

1.5 Prerequisites to Analysis

Our approach to analysis is directed by the theoretical context and the experimental design. This book is not about design, but we must be aware that design impacts analysis. We can identify two extreme ways of solving the design problem [4]:

EXTENSIVE DESIGN, in which we measure many samples then search for patterns and typical behavior by some kind of averaging. The objective is to obtain highly precise results.

INTENSIVE DESIGN, in which we measure only a few samples, but do so as thoroughly and carefully as possible. The objective here is to obtain highly accurate results.

Of course, these are only the extremes; in practice, we may seek a middle ground between precision and accuracy. In any case, success requires us to know the levels of precision and accuracy we hope to achieve, then design and analyze accordingly.

The inexperienced may tend to view data in a detached way, as a collection of numbers extracted and removed from the experiment that produced them. This can hamper analysis for such a detachment can promote the attitude that precision and accuracy are absolutes, not subject to interpretation or revision. But a complete and insightful analysis of data demands that we [5]

- Understand how the apparatus works,
- Understand why particular techniques were chosen for obtaining data from the apparatus,
- Know the limitations of the techniques and the apparatus,
- Understand the theoretical context.

Without such knowledge, our analyses will be, at best, superficial interpretations of the data. More likely, they will be wrong.

Undoubtedly, books such as this one contribute to the tendency of separating data from experiment, for in this book we repeatedly offer examples using numbers divorced from an experimental context. Our situation is much like that of an athlete or performing artist. Part of a musician's training involves practicing scales, even though a complete scale rarely appears in a piece of music. So why practice such activities? Because they are fundamental. They exercise skills: they train the hand, they train the ear, they train the mind.

The following chapters can serve a similar function: they are fundamental; they provide opportunities to practice skills; they can help train the mind. They are not, however, a complete symphony.

Literature Cited

[1] P. Duhem, *The Aim and Structure of Physical Theory*, translated by P. P. Wiener, Princeton University Press, Princeton, NJ, 1954.

[2] P. J. Galison, *How Experiments End*, University of Chicago Press, Chicago, 1987.

[3] G. Polya, *Mathematics and Plausible Reasoning*, Vol. 1, "Induction and Analogy in Mathematics," Princeton University Press, Princeton, NJ, 1954, p. 204.

[4] R. Harré, *Great Scientific Experiments*, Oxford University Press, Oxford, 1981, p. 193.

[5] R. Root-Bernstein, *Discovering*, Harvard University Press, Cambridge, MA, 1989, p. 252.

Exercises

1.1 Assign upper and lower bounds to each of the following and explain the bounds you choose:

(a) $\sqrt{19354}$

(b) The density of gasoline

(c) The vapor pressure of water at 70°C

(d) The number of drug stores in Wyoming

(e) The viscosity of pancake syrup at room temperature

1.2 Make a reasoned guess for each of the following values and explain your reasoning:

(a) The number of words in this book

(b) The diameter of the earth's moon

(c) The thickness of one page of this book

1.3 List the assumptions you would make that would allow you to estimate each of the following. No calculations are required.

(a) The density of air at ambient conditions

(b) The diameter of one water molecule

(c) Energy used in playing an audio CD for one hour

1.4 Describe, in as much detail as you can, a measurement or sequence of measurements that would enable you to estimate each of the following:

(a) The mass flow rate of water dripping from a faucet

(b) The height of a tree

(c) The power (watts) provided by a microwave oven

(d) The efficiency of a 100-watt incandescent light bulb. Efficiency refers to the fraction of electrical energy supplied to the bulb that is converted into visible light.

Exhibit B

The gravitational force between an object of mass M and another of mass m is given by $F = GMm/r^2$, where r is the distance between their centers and G is the universal gravitational constant—one of the fundamental constants of nature. In 1986 the Committee on Data for Science and Technology (CODATA) of the International Council of Scientific Unions established the value of G with an uncertainty of 0.013% [1]. At the end of 1999 the Committee took the unusual step of increasing the assigned uncertainty to 0.15% [2]. Here are some of the values for G measured between 1990 and 2000.

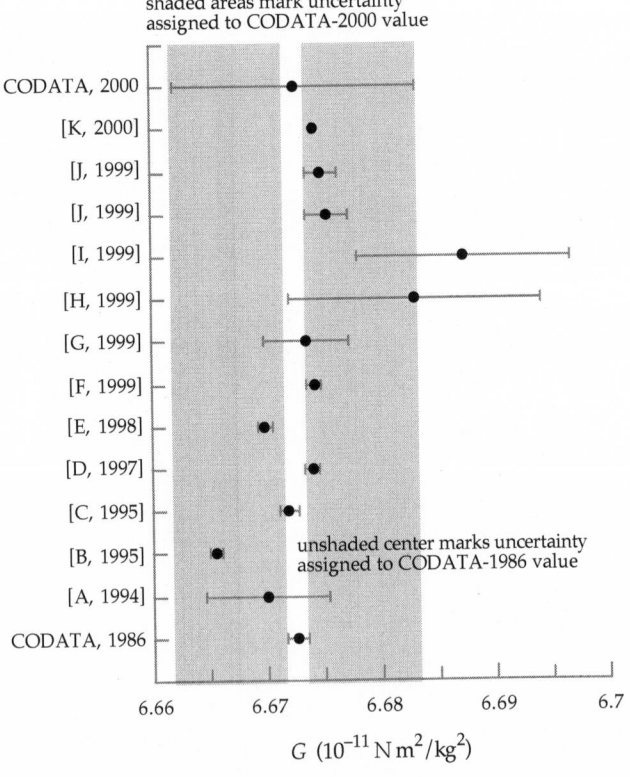

Citations for references A–K are listed in Appendix A.

2

Assessing the Quality of Measurements

N O MEASUREMENT IS EXACT. Every measurement has some error associated with it; so, the exact value for any measured quantity is unknown. Likewise, the error is unknown, for to know the error requires us to know the exact value. We can distinguish between kinds of errors, as in § 2.1, but what we can actually quantify are *uncertainties*—measures of credibility. Therefore, this chapter focuses on uncertainties: how to estimate them, how to combine them, and how to report them. Our presentation follows the guide to uncertainties adopted by the American National Standards Institute [3].

2.1 Errors

Let y represent the (unknown) exact value of a measurable quantity and let y_i represent one value measured for y. Then the error e_i in y_i is the difference

$$e_i = y_i - y \tag{2.1}$$

The error divides in two: *systematic* error and *statistical* error.

Systematic errors detract from the *accuracy* of measurements, statistical errors detract from the *precision* of measurements. Accuracy and precision are mutually independent, so we can identify four possibilities: (a) low accuracy and low precision, (b) low accuracy and high precision, (c) high accuracy and low precision, (d) high accuracy and high precision. These are illustrated in Figure 2.1.

Systematic error is the difference between the (unknown) exact value y and the mean y_m of an infinite number of measurements,

$$e_{sys} = y_m - y \tag{2.2}$$

Systematic errors are caused by some consistent bias in the measurement, such as a faulty instrument calibration, impurities in samples, consistent misreading of a meter. Such errors cannot be determined by

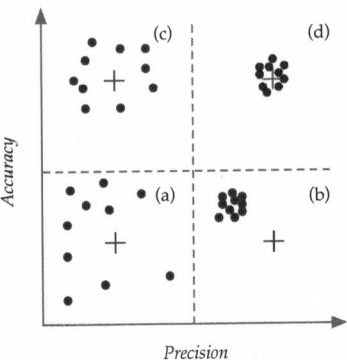

FIGURE 2.1 The cross in each panel represents the exact value of a quantity that is to be measured. When the quantity is measured ten times, each panel illustrates one of four possible outcomes. The ten measurements could be (a) low in both accuracy and precision, (b) low in accuracy but high in precision, (c) high in accuracy but low in precision, or (d) high in both accuracy and precision.

mathematical manipulations applied to measurements repeated at one x-value. Rather, estimates may be obtained by (a) thoughtfully considering how bias might enter a measurement and by (b) comparing measured values to other, independent experimental or theoretical results. When systematic errors can be identified and estimated, we might legitimately try to compensate for them by applying correction factors to measured values.

Design and execution of experiments focus on eliminating or controlling systematic error; that is, a good experiment pushes systematic error into the background, so that the phenomena of interest are apparent and accessible [4]. This becomes more challenging as we move to higher accuracy: not only must we control large effects more tightly, but subtle effects may become important when high accuracy is sought. To paraphrase Galison [4], *some systematic errors leave light footprints*.

Statistical (also called random) error is the difference between one measured value and the mean computed from an infinite number of repeated measurements,

$$e_{sta} = y_i - y_m \tag{2.3}$$

These errors are caused by momentary bias or erratic fluctuations, such as a voltage surge, a pressure fluctuation, or a lapse in reading a meter.

Note the following: (a) The sum of (2.2) and (2.3) gives the total error (2.1). (b) It is not the source of an error, but its regularity that dictates whether the effect is systematic or statistical. (c) We do not know the exact value y nor can we perform an infinite number of measurements to obtain the mean y_m; hence, systematic and statistical errors are both unknown and unknowable.

2.2 The Gaussian Distribution

To explore the range over which a measured value is distributed, we repeat a measurement several times. When fluctuations in the measurement are dominated by random events that are mutually independent, repeated values measured for y describe a *Gaussian* distribution [5] about their mean value y_m:

$$f(y) \;=\; \frac{1}{s\sqrt{2\pi}} \exp\left[\frac{-(y-y_m)^2}{2s^2}\right] \tag{2.4}$$

This is sometimes called the *normal* distribution.

In (2.4) the quantity s is called the *standard deviation*; for N repeated measurements, y_i ($i = 1, 2, \ldots, N$), it is computed by

$$s \;=\; \frac{\sqrt{\sum_i^N (y_i - y_m)^2}}{\sqrt{N-1}} \;\propto\; \frac{\text{noise}}{\text{signal strength}} \tag{2.5}$$

It measures the width of the distribution $f(y)$, as in Figure 2.2. Note

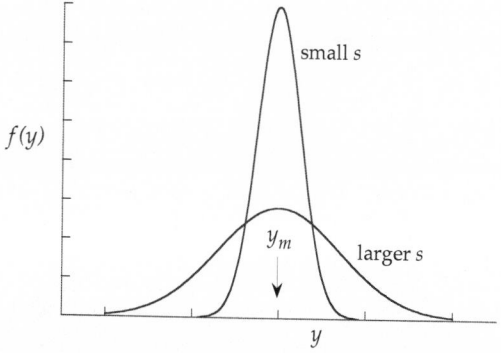

FIGURE 2.2 The Gaussian distribution (2.4) for two values of the standard deviation s. When s is small, most measured values of y lie close to their mean y_m; when s is large, more values lie farther away.

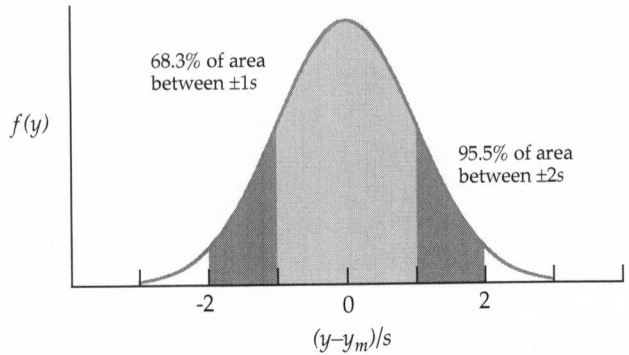

FIGURE 2.3 When repeated measurements obey a Gaussian distribution, then 68.3% of the values lie within ±1s of the mean and 95.5% lie within ±2s.

that s is inversely proportional to the signal-to-noise ratio: a large standard deviation implies a relatively large amount of noise with measured values widely scattered about their mean. In such cases, we attempt to increase the signal-to-noise ratio by increasing N. Note that for one measurement ($N = 1$), (2.5) gives $s = 0/0$; that is, a single measurement gives us no information about the width of the distribution. If we integrate portions of the Gaussian, we obtain the number of values that lie within a specified distance from the mean. In particular, we find that, as in Figure 2.3,

- 68.3% of the measured values lie within ±1s of y_m,
- 95.5% lie within ±2s of y_m, and
- 99.7% lie within ±3s of y_m.

In Figure 2.3 the maximum in the Gaussian distribution occurs at the mean value of y; that is, the mean is the most probable value—the value most likely to be measured.

2.3   Exploring Repeated Measurements

We will use repeated measurements to help us quantify uncertainties (§ 2.4), but other information can also be extracted from repeated measurements. The activities introduced here are simple but they are not unimportant [6]. We illustrate using an example.

Table 2.1 Ten measured values for the same volumetric flow rate y of water through a pipe. These ten values have a mean $y_m = 5.67$ gpm. Included here are the deviations and squares of deviations from that mean.

Run #	y (gpm)	δy (gpm)	$(\delta y)^2$ (gpm)^2
1	5.5	−0.17	0.029
2	5.85	0.18	0.032
3	5.55	−0.12	0.014
4	5.8	0.13	0.017
5	5.9	0.23	0.052
6	5.6	−0.07	0.005
7	5.75	0.08	0.006
8	5.65	−0.02	0.0004
9	5.4	−0.27	0.073
10	5.7	0.03	0.0009

EXAMPLE: The steady flow of water through a pipe has been measured ten times; the results in gallons per minute (gpm) are given in Table 2.1. We explore these data by determining three characteristics: central values, variability, and distribution.

2.3.1 Central Values

Central values guide our thinking about the magnitudes of individual measurements. Two kinds of central values are common: the *mean* and the *median*. The mean is computed by simple averaging; for the data in Table 2.1 this gives

$$\text{mean} = y_m = \frac{1}{N} \sum_i^N y_i = 5.67 \text{ gpm} \tag{2.6}$$

To find the median, we list the values in ascending or descending order:

$$5.4, \ 5.5, \ 5.55, \ 5.6, \ 5.65, \ 5.7, \ 5.75, \ 5.8, \ 5.85, \ 5.9$$

For an odd number of measurements, the median is the one in the middle. For an even number, it is the average of the two in the middle. So for our ten values, we find

$$\text{median} = (5.65 + 5.7)/2 = 5.68 \, \text{gpm} \qquad (2.7)$$

This median is close to the mean, suggesting that the distribution is roughly symmetric about its center. If the two had not been close, then the distribution would be asymmetric.

Note that the median is less sensitive to small changes in the measurements than is the mean. For example, if the largest value had been 10.0 rather than 5.9, the median would not have changed, but the mean would have increased. For this reason, the median is said to be a more *robust* measure of the center than the mean.

2.3.2 Variability

We have two simple measures of variability; one is the range, which is the difference between the largest and smallest values:

$$\text{range} = 5.9 - 5.4 = 0.5 \, \text{gpm} \qquad (2.8)$$

A narrow range centered near the median suggests precise measurements symmetrically distributed about a central value. The data in Table 2.1 have a range that is about 9% of the median, and the range is centered at 5.65 gpm, which is near the median at 5.68 gpm.

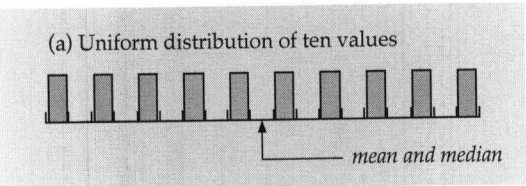

(a) Uniform distribution of ten values

mean and median

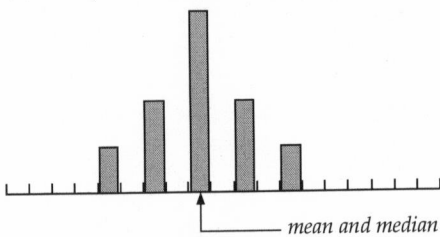

(b) Nearly normal distribution of ten values

mean and median

FIGURE 2.4 Examples of common frequency distributions for ten repeated measurements of the same quantity. In both uniform (a) and normal (b) distributions, the mean and median coincide.

Variability is also measured by the standard deviation (2.5). For our flow example, calculation of the standard deviation is outlined in Table 2.1, which contains the deviation from the mean and squared deviation for each measured value. Note that the deviations sum to zero, while the squared deviations sum to 0.23. Then from (2.5), we find $s = (0.23/9)^{1/2} = 0.16$ gpm, which is 3% of the mean.

The range is a more robust measure of variability than is the standard deviation; however, the standard deviation provides more information. The standard deviation measures the dispersion of values about their mean: a small s implies clustering of values near the mean. Because contributions to the sum in s are squared (2.5), the value for s is strongly influenced by those data that fall far from the mean.

2.3.3 Frequency Distribution

A third characteristic is the frequency distribution for repeated measurements: a histogram for the number of times each value was measured. We want to know the general shape of this distribution; that is, is it roughly normal or uniform, as in Figure 2.4, or is it clustered or skewed, as in Figure 2.5? We also want to identify possible *outliers*: values that seem to be far removed from the others. Outliers may be artifacts—values caused by some blunder in the measurement. Or they

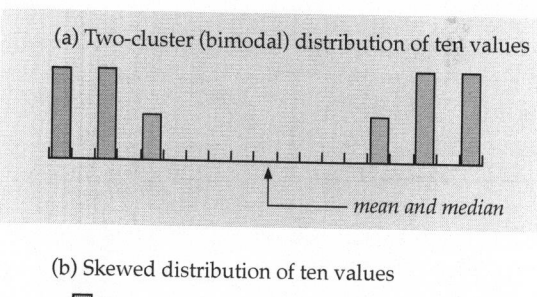

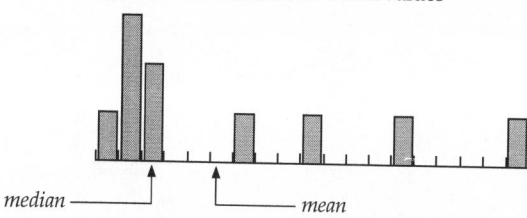

FIGURE 2.5 More examples of common frequency distributions for ten repeated measurements of the same quantity. In bimodal (a) and skewed (b) distributions the mean and median may, or may not, coincide.

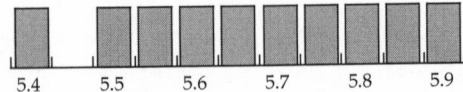

FIGURE 2.6 Frequency distribution for the flow rates in Table 2.1, showing a nearly uniform distribution with mean = 5.67, median = 5.68, and a possible outlier at 5.4.

may be real, caused by some unsuspected phenomenon or by an unexpected sensitivity to some variable.

For the flow rates in Table 2.1, the distribution is roughly uniform with a possible outlier, as shown in Figure 2.6. To check whether the point at 5.4 is indeed an outlier, we apply the *one-point test:* does the removal of one point significantly change our impression of the distribution? Ignoring the point at 5.4, the mean of the nine remaining points is 5.7 and their median is also 5.7: the mean, median, and overall distribution are little affected. This suggests that the point is not an outlier; a statistical test for outliers is discussed in § 2.3.4.

Note that the shape of a frequency distribution is influenced by the width of the bins used to construct the histogram. In other words, it is affected by the number of significant figures used for the data. For example, the histogram shown above is based on three significant figures for the data in Table 2.1. But if we round to two significant figures, then the frequency distribution appears as in Figure 2.7. Figures 2.6 and 2.7 together illustrate *coarse-graining,* and we must be aware that our interpretations are influenced by the scale at which we make observations.

We have now presented three simple tests to apply to repeated measurements of the same quantity under the same apparent conditions: central values (mean and median), variability (range and standard deviation), and frequency distribution. These simple tests, and the conclusions we might draw from them, are summarized in Table 2.2.

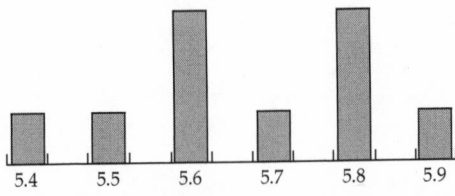

FIGURE 2.7 Frequency distribution for the flow rates in Table 2.1, but with values rounded to two significant figures. Compare this with the distribution in Figure 2.6 for the same data, but there rounded to three significant figures.

TABLE 2.2 Tests to apply to repeated measurements.

Test	Implication
Compare median with mean	Median not near mean suggests strong asymmetry in the distribution.
Compare median with range	Small range suggests high precision, large range suggests low precision.
Compare standard deviation with mean	Small standard deviation suggests a tight clustering about mean.
Compare standard deviation with range	Range larger than standard deviation suggests presence of an outlier.
Plot and identify the distribution	Strong clustering or skewedness suggests something changed during the measurements.
Identify any outliers	Outliers might be caused by blunders in performing some measurements.

2.3.4 Discarding Data

When we repeat measurements of the same quantity, we sometimes obtain one value that differs noticeably from the others. We have an outlier and the question is, what should we do about such a value? Under what conditions might we discard that measurement?

On the one hand, an outlier might result from some blunder in performing the measurement: we misread a meter, or an instrument drifted out of calibration, or a thunderstorm caused a spurious voltage fluctuation. On the other hand, an outlier might signal the presence of some unexpected phenomena. An example is Lord Rayleigh's preoccupation with small discrepancies in his measured masses of volumes of air, which led him to discover argon [7]. Outliers may be unexpected, but they are neither uncommon nor necessarily unimportant [8].

The issue is what to do when you cannot decide whether an outlier is either a blunder or a real phenomenon. One way to distinguish between these two possibilities is to apply *Chauvenet's criterion* [5]: if the outlier lies more than t standard deviations from the mean of N measurements, then it is probably not a natural fluctuation, so you may consider discarding that value. The value of t depends on N; minimum values of t for various N are given in Table 2.3.

Note that we have only *suggested* that you discard the measurement; it would be better to keep the value, but repeat the measurements

TABLE 2.3 Lower bounds on t, the number of standard deviations above which we might discard an outlier from N measured values.

N	t	N	t
3	1.38	16	2.16
4	1.54	18	2.20
5	1.65	20	2.24
6	1.74	25	2.33
8	1.86	30	2.39
10	1.96	40	2.50
12	2.04	50	2.58
14	2.10	100	2.81

in an attempt to explain it or to reduce its impact on the mean and standard deviation. In practice, however, constraints of time and resources often prevent additional experiments.

To illustrate, we reconsider the ten measured flow rates from Table 2.1. The mean of those ten values is 5.67 gpm, with a possible outlier at 5.4 gpm. To apply Chauvenet's test, we compute t,

$$t = \frac{|y - y_m|}{s} \qquad (2.9)$$

where y is the value of the possible outlier, y_m is the mean of the N measurements, and s is the standard deviation of any one value from the mean. For our example, we find

$$t = \frac{|5.4 - 5.67|}{0.16} = 1.7 \qquad (2.10)$$

But from Table 2.3, we find that for $N = 10$, we must have $t \geq 1.96$ before we could consider discarding a measured value. This means outliers are expected to occur at $x < 5.36$ and $x > 5.98$. So in this case, we judge that the measurement at 5.4 gpm is not an outlier.

Chauvenet's criterion is a purely statistical test. More often decisions about discarding or keeping data are made within the theoretical context that guides the experiment. The theoretical context may identify some systematic error that explains an outlier and allows us to remove it from further analysis. We emphasize that considered judgments to discard data do not necessarily signal poor experimental procedures; rather, such judgments are a legitimate part of any experimental process [4].

But no matter what we decide, we must report *all* measured data and then discuss what steps were taken to explain possible outliers. If any data are discarded, we must still report those values and explain the criteria used to justify eliminating particular points from subsequent analyses. In this way we maintain a complete document of what was done; so, if more information becomes available at some later date, we might be able to explain or reconcile outliers.

2.4 Uncertainties

Although errors cannot be known, it is possible to quantify uncertainties; an *uncertainty* is a range that probably contains the (unknown) exact value. This range is consistent with the measured data, the experimental protocol, and our understanding of how the world works. Following the ANSI guide [3], we divide uncertainties into two kinds:

1. Type A uncertainties are those in which we use repeated sampling to quantify the distribution of values. We then estimate the uncertainty by performing a statistical analysis on the repeated measurements.

2. Type B uncertainties are those in which we assume a distribution based on experience, other experiments, reference works, etc. Consequently, we do not use a statistical analysis to estimate these uncertainties.

We caution that Type A uncertainties are not necessarily confined to statistical errors nor are Type B uncertainties necessarily limited to systematic errors. Uncertainties are independent of errors; we do not attempt to quantify errors, which are unquantifiable, but we do quantify uncertainties.

2.4.1 Evaluating Type A Uncertainties

We illustrate by continuing to analyze the flow-rate measurements in Table 2.1. In those measurements, the sources of uncertainty include any fluctuations in the behavior of the meter, in the way we read it, and in the flow itself. We treat these fluctuations as Type A uncertainties and estimate their combined effects by repeating the measurement ten times. We have already found the mean of those to be 5.67 gpm and their standard deviation to be $s = 0.16$ gpm.

This standard deviation s applies to any one measurement of y. It means that if we were to make additional measurements, then we expect that, from Figure 2.3, 68.3% of them would fall within $\pm 1s$ of our mean and 95.5% would fall within $\pm 2s$. The values 68.3% and 95.5% indicate how confident we would be about a particular measurement; hence, they are called *confidence levels*.

But rather than report any one measured value of y, the mean y_m would be more reliable. And as a measure of the quality of the mean, computed from N measured values of y, we use its standard deviation s_m. From the propagation of uncertainties procedure in § 2.5, we can show that s_m is related to s by [5]

$$s_m = \frac{|s|}{\sqrt{N}} \tag{2.11}$$

For our $N = 10$ measurements, (2.11) gives

$$s_m = \frac{0.16}{\sqrt{10}} = 0.051 \text{ gpm} \tag{2.12}$$

We use this standard deviation as an estimate of the uncertainties of Type A,

$$u_A = s_m \tag{2.13}$$

2.4.2 Evaluating Type B Uncertainties

For those uncertainties that cannot be estimated from repeated measurements, we appeal to calibration standards, to manufacturer's specifications, to experience with the equipment, to knowledge of physical properties, to judgement about factors that cause variations in measured values, and the like.

For the flow experiment of Table 2.1, we identify three sources of uncertainty that we treat as Type B.

(a) The flow meter was calibrated by the manufacturer at 21°C with an uncertainty of $\pm 2\%$ of scale reading. For a mean flow rate of 5.67 gpm, this gives an uncertainty of ± 0.113 gpm.

(b) Our measurements were not done at 21°C, so we include uncertainties caused by variations in fluid temperature. We estimate the temperature might vary by as much as $\pm 2^\circ$C; but from NIST tables for water [9], even a $\pm 5^\circ$C change at 20°C causes only 0.1% change in the molar volume of water. This gives a variation of ± 0.006 gpm in the flow. This will probably be negligible compared to other uncertainties.

(c) The scale resolution on the meter is 0.1 gpm. We believe we can read the meter to within half of this, so we estimate that meter readings may be uncertain up to ± 0.025 gpm.

This example is typical: Type B uncertainties are often composed of several independent contributions, u_{Bi}. The total u_B is computed from the individual components by [3, 5],

$$u_B = \sqrt{\sum u_{Bi}^2} \qquad (2.14)$$

In our flow-rate problem we identified three Type B uncertainties: one of 0.113 gpm, one of 0.006 gpm, and one of 0.025 gpm. Hence,

$$u_B = \sqrt{0.113^2 + 0.006^2 + 0.025^2} = 0.12 \text{ gpm} \qquad (2.15)$$

Often we are not confident that we can identify all contributions to Type B uncertainties. But if, based on our knowledge and experience, we can bound the expected value, then we can still assign a value to u_B. Here are two examples [3].

50/50 BOUND. Perhaps we are *certain* that the (unknown) exact value for y has a 50/50 chance of lying between a lower bound y_{lo} and an upper bound y_{hi}. Further, we believe values for y are normally distributed on this range. Then let a be the half-width of the interval,

$$a = \tfrac{1}{2}(y_{hi} - y_{lo}) \qquad (2.16)$$

and take the Type B uncertainty to be

$$u_B = 1.48\,a \qquad (2.17)$$

because the range $[-a/1.48, a/1.48]$ captures 50% of a normal distribution. Obviously, this procedure can be generalized to bounds having other probabilities of occurring. For example, if the interval $2a$ has a two out of three chance of bounding the desired value, then the factor in (2.17) becomes 1 rather than 1.48.

BOUNDS ONLY. In other cases we may only be able to bound the value. That is, we are certain the (unknown) exact value for y lies between y_{lo} and y_{hi}, but we do not know the distribution. Then we can only assume a uniform distribution: every value in the range is equally likely to occur. In this case, take the Type B uncertainty to be the standard deviation for the midpoint of the range $[y_{lo}, y_{hi}]$; that is, use

$$u_B = \frac{a\sqrt{3}}{3} \qquad (2.18)$$

where a is, again, the half-width (2.16).

2.4.3 Combined Uncertainty

For a single measured quantity y, the total or combined uncertainty u accumulates Type A and Type B uncertainties. We assume Type A and Type B uncertainties are independent, so the total uncertainty is computed analogously to (2.14) [3, 5]; see Figure 2.8,

$$u = \sqrt{u_A^2 + u_B^2} \qquad (2.19)$$

To complete our flow-rate example, we have the Type A uncertainty from (2.13) as $u_A = 0.051$ gpm and we found the Type B uncertainty in (2.15) to be $u_B = 0.12$ gpm. So (2.19) gives

$$u = \sqrt{0.051^2 + 0.12^2} = 0.13 \text{ gpm} \qquad (2.20)$$

When reporting a total uncertainty, we usually keep one significant

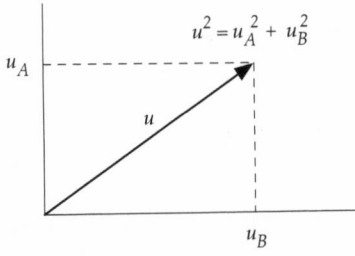

FIGURE 2.8 The total (combined) uncertainty is the root of the sum of squares of the Type A and Type B uncertainties, as in (2.19).

figure; however, an exception occurs when the first significant digit is unity. Then we may include one additional figure, as in (2.20).

2.4.4 Expanded Uncertainty

In many situations we want to be able to state that some interval captures a large fraction of the values that could be measured for the quantity; that is, we need to report an *expanded* uncertainty U to which we can attach a level of confidence. This expanded uncertainty is computed from our total uncertainty u by multiplying by a coverage factor k [3],

$$U = ku \qquad (2.21)$$

We commonly take the coverage factor from a normal distribution, unless some other distribution is known to apply. Therefore, we use

- $k = 1$ to claim a level of confidence of 68.3%,
- $k = 2$ to claim 95.5%,
- $k = 2.58$ to claim 99%, and
- $k = 3$ to claim 99.7%.

For Type A uncertainties, coverage factors from the normal distribution can usually be used when the number of measurements N is large. But when the number is small (say, $N < 12$), the sampling distribution is ill-defined and the coverage factor from the normal distribution is too small. In such cases, we separate the Type A coverage factor k_A from the Type B factor k_B and replace (2.21) with

$$U = \sqrt{(k_A u_A)^2 + (k_B u_B)^2} \qquad (2.22)$$

Values for k_B are still taken from a normal distribution, but values for k_A are now taken from Table 2.4 at the 68%, 95%, and 99% levels of confidence. Note that as N increases, the values of k_A in Table 2.4 approach those for a normal distribution; that is, as N increases, $k_A \rightarrow k_B$ and (2.22) reverts to (2.21).

We can now estimate a value for the expanded uncertainty of the ten flow rates given in Table 2.1. If we choose the 95% confidence level, then U is given by (2.22) as

$$U = \sqrt{(2.2 \times 0.051)^2 + (1.96 \times 0.12)^2} = 0.3 \text{ gpm} \qquad (2.23)$$

TABLE 2.4 Values of Type A coverage factor k_A for N measurements at three levels of confidence [10].

N	68%	95%	99%
∞	1	1.96	2.6
20	1.06	2.1	2.8
10	1.09	2.2	3.2
8	1.11	2.3	3.4
6	1.13	2.4	3.7
4	1.2	2.8	4.6
3	1.25	3.2	5.8
2	1.39	4.3	9.9

Note that the value of 1.96 for k_B is also taken from Table 2.4: it appears for an infinite number of samples, for which we assume the measurements would be distributed as a Gaussian about their mean. Finally, we report the measured flow rate, at the 95% confidence level, as

$$y = 5.7 \pm 0.3 \text{ gpm} \tag{2.24}$$

The position of the one significant figure in the uncertainty (0.3) determines the position of the last significant figure in the reported mean (5.7). In contrast, if we reported a mean of 5.70, we would be exaggerating the quality of the result because the final "0" is, in fact, insignificant compared to an uncertainty of ± 0.3 gpm. We caution that the $\pm$ notation should be applied only to expanded uncertainties; see § 2.6.

2.5 Propagation of Uncertainties

In analyzing experimental data we routinely combine measured values to compute values for other quantities. For example, from values measured for the temperature T, pressure P, volume V, and number of moles n of a gas sample we might compute a value for the compressibility factor Z,

$$Z = \frac{PV}{nRT} \tag{2.25}$$

Each measured quantity has some uncertainty, and so the question is this: How do those uncertainties combine to yield the uncertainty in the value for a computed quantity [3, 5]? This problem is called the *propagation of uncertainties*; here we first present the general expression (§ 2.5.1) and then we consider common special cases (§ 2.5.2) for determining uncertainties in computed quantities.

2.5.1 General Expression

Consider a quantity y whose value is to be computed from values measured for other quantities $x_1, x_2, \ldots$.

$$y = f(x_1, x_2, \ldots) \tag{2.26}$$

This equation is commonly called a *measurement* equation; it is usually derived from a model of the experimental situation. Each measured x_i has some uncertainty $u_{xi} > 0$. We expand the (unknown) exact value of y in a Taylor series about the computed value, keeping only the first-order terms,

$$y = y_{cal} + \sum_i \left(\frac{\partial f}{\partial x_i} \right) u_{xi} \tag{2.27}$$

The partial derivatives measure the sensitivity of y to changes in the xs, as discussed in Chapter 5. We now estimate the uncertainty in y as

$$u_y = |y - y_{cal}| \approx \sum_i \left| \left(\frac{\partial f}{\partial x_i} \right) \right| u_{xi} \tag{2.28}$$

Recall that uncertainties, such as u_{xi} and u_y, are always positive.

2.5.2 Special Cases

When y is obtained from simple sums or differences of the measured xs, then (2.28) gives the uncertainty in y as the algebraic sum of the individual uncertainties. For example, say we have measured the length ℓ and width w of a rectangle; the measured values have uncertainties u_ℓ and u_w. Then the uncertainty in the computed perimeter p, from (2.28), is given by

$$u_p = 2u_\ell + 2u_w \tag{2.29}$$

Terms on the rhs of an expression like (2.29) *always* add because uncertainties are always positive and because absolute values are used for derivatives in (2.28).

When y is obtained from a product or quotient of measured xs, then (2.28) gives the fractional uncertainty in y as the sum of the fractional uncertainties in the xs. For example, the uncertainty in the rectangular area A computed from a measured length and width would be

$$\frac{u_A}{A} = \frac{u_\ell}{\ell} + \frac{u_w}{w} \qquad (2.30)$$

Likewise, for our example of the compressibility factor (2.25), the uncertainty in Z would be obtained by

$$\frac{u_Z}{Z} = \frac{u_P}{P} + \frac{u_V}{V} + \frac{u_n}{n} + \frac{u_T}{T} \qquad (2.31)$$

When y involves other functions of the xs (such as powers, logs, exponentials, etc.), then merely apply (2.28) to obtain the required expression for the uncertainty. For example, the volume of a right cylinder is

$$V = \pi r^2 h \qquad (2.32)$$

So for measured values of the radius and height, the uncertainty in the computed volume, from (2.28), is given by

$$u_V = (2\pi r h)u_r + (\pi r^2)u_h \qquad (2.33)$$

On dividing this through by V, from (2.32), we obtain

$$\frac{u_V}{V} = \frac{2u_r}{r} + \frac{u_h}{h} \qquad (2.34)$$

Besides experimental analysis, expressions for uncertainties in computed quantities can also be used in experimental design. For example, assume we need to design an experiment for obtaining Z for a gas by measuring $P, V, n,$ and T. Equation 2.31 shows that the fractional uncertainty in Z is obtained by accumulating the fractional uncertainties in the measured quantities. Therefore, at fixed T and P, (2.31) suggests that we might reduce the fractional uncertainty in Z by placing a large number of moles n of the gas in a vessel having a large volume V. Further, in an expression like (2.31), some terms usually make larger contributions than others. For example, we can often measure $n, P,$ and T more accurately than V; in such cases, the uncertainty in a value computed for Z may be dominated by the uncertainty in V and then those in $n, P,$ and T can be neglected.

TABLE 2.5 Uncertainty Budget for Water-Flow Example Developed in § 2.4.

Source of Uncertainty	Type	Uncertainty (gpm)
Calibration curve at 21°C	B	0.113
Fluctutations in flow and meter	A	0.051
Scale resolution of meter	B	0.025
Variations in fluid temperature	B	0.006
Total (combined) uncertainty	...	0.13

2.6 Reporting Uncertainties

When reporting how uncertainties were estimated, we should cite the source of each, tell whether it was treated as Type A or Type B, and give the magnitude of each estimate. This can be conveniently done in a table, called an *Uncertainty Budget* [11]; to illustrate, the uncertainty budget from our analysis of the flow-rate measurements is given in Table 2.5. Note that an uncertainty budget can be a useful tool in experimental design as well as in experimental analysis. During design an uncertainty budget could help us identify possible sources of uncertainties and assess the relative importance of each.

When reporting final values for uncertainties, we should strive for consistency by using standard formats that help readers understand exactly what is being reported. The preferred ways for reporting final values of uncertainties are given in the ANSI guide [3] and are summarized in Table 2.6. Note the following about entries in Table 2.6:

(a) For the total (combined) uncertainty we may use either of two formats: format 1 [$y = 5.7$ gpm with 0.1 gpm total uncertainty] or format 2 [$y = 5.7(1)$ gpm].

(b) In format 2 [$y = 5.7(1)$ gpm] of Table 2.6, the integer in parenthesis is the combined uncertainty for the last digit in the mean.

(c) When reporting expanded uncertainties, be sure to include the level of confidence adopted. Typically, we use either a 68%, 95%, or 98% confidence level.

(d) The $\pm$ notation is reserved for reporting expanded uncertainties. This may be done as in format 4 [$y = 5.7 \pm 0.3$ gpm] or in format 5 [$y = 5.7$ gpm $\pm 5\%$].

TABLE 2.6 Recommended Formats for Reporting Total and Expanded Uncertainties [3]. Note the two formats for reporting a total uncertainty.

Uncertainty	Reporting Format
Total	1. $y = 5.7$ gpm with 0.1 gpm total uncertainty
	2. $y = 5.7(1)$ gpm
Relative Total	3. $y = 5.7$ gpm with 2% total uncertainty
Expanded	4. $y = 5.7 \pm 0.3$ gpm at the 95% confidence level
Relative Expanded	5. $y = 5.7 \pm 5\%$ at the 95% confidence level

2.7 Summary of Steps for Evaluating Uncertainties

The procedures discussed in § 2.3–2.6 require considerable thought and planning. To help organize your thinking, here is a summary of the steps that should be followed in evaluating and reporting uncertainties in measured quantities:

1. Perform repeated measurements and use them as the basis for obtaining a mean and estimating the Type A uncertainty; see § 2.3 and § 2.4.1.

2. Use your knowledge of the apparatus, calibration curves, effects of the environment, etc. to estimate the type B uncertainty; for an example, see § 2.4.2.

3. Use the procedure in § 2.4.3 to combine your values for the Type A and Type B uncertainties to obtain the total uncertainty.

4. If you want to assign a level of confidence, apply a coverage factor to the total uncertainty; this yields the expanded uncertainty, as shown in § 2.4.4.

5. If you use measured values to compute other quantities, use the propagation of uncertainties procedure from § 2.5 to find the uncertainties in those computed results.

6. Use an uncertainty budget, as in Table 2.5, to show how uncertainties were determined.

7. Use a standard format from Table 2.6 to report final values for means and their uncertainties.

When constructing x-y plots, we commonly include an "error bar" at each point; the values represented by such "error bars" should be the expanded uncertainties. Moreover, the level of confidence chosen to determine those expanded uncertainties should be stated in the figure caption. Therefore, the term "error bar" is, in fact, a misnomer; it should be called an *uncertainty bar*.

Every measured quantity has some uncertainty associated with it, and so, by attaching an uncertainty to every reported quantity, you can literally double the amount of useful information extracted from an experiment. Since measurements require time, effort, and resources, to neglect the evaluation of uncertainties is, at the least, uneconomical.

Literature Cited

[1] E. R. Cohen and B. N. Taylor, "The 1986 Adjustment of the Fundamental Physical Constants," *Rev. Mod. Phys.*, 59, 1121 (1987); first published as CODATA Bulletin No. 63, Pergamon, Oxford and New York, Nov., 1986.

[2] P. J. Mohr and B. N. Taylor, "Adjusting the Values of the Fundamental Constants," *Phys. Today*, 54, 29 (3/2001).

[3] *U. S. Guide to Expression of Uncertainty in Measurement*, American National Standards Institute, NCSL International at Boulder, CO, 1997; ANSI/NCSL Z540-2-1997.

[4] P. J. Galison, *How Experiments End*, University of Chicago Press, Chicago, 1987.

[5] J. R. Taylor, *An Introduction to Error Analysis*, University Science Books, Mill Valley, CA, 1982.

[6] J. W. Tukey, *Exploratory Data Analysis*, Addison-Wesley, Reading, MA, 1977.

[7] Lord Rayleigh, *Proceedings of the Royal Society*, 55, 340 (1894); see also *Scientific Papers by Lord Rayleigh*, Dover, New York, vol. IV, 104 (1964).

[8] N. N. Taleb, *The Black Swan: The Impact of the Highly Improbable*, Random House, New York, 2007.

[9] W. G. Mallard and P. J. Linstrom, eds., *Chemistry Webbook, NIST Standard Reference Database Number 69*, Feb. 2000, Gaithersburg, MD. Website at http://webbook.nist.gov/chemistry/.

[10] B. W. Lindgren and G. W. McElrath, *Introduction to Probability and Statistics*, Macmillan, New York, 1959.

[11] B. N. Taylor and C. Kuyatt, "Guidelines for Evaluating and Expressing the Uncertainty of NIST Measurement Results," *NIST Technical Note 1297*, U. S. Government Printing Office, Washington, DC, 1994.

Exercises

2.1   Indicate whether each of the following affects accuracy or precision of measured data.

(a) Events are timed with a stop watch that is slow by 2 s.

(b) Measurements of water flow rate through a pipe change with the flushing of a toilet in a nearby rest room.

(c) Fluid volumes are measured in a cylinder whose walls expand with increasing ambient temperature.

(d) Measurements of surface tension change as the laboratory vibrates in response to traffic along an adjacent road.

2.2   In (2.22) the value of the coverage factor for the Type A uncertainty differs from that for the Type B uncertainty. Why?

2.3   The linear velocity of water flowing through a pipe is to be determined by a bucket-and-scales method. The pipe diameter is measured to be 2.5 ± 0.1 cm. For a duration of 1.00 ± 0.02 min the amount of water collected in the bucket is found to be 42.30 ± 0.08 kg. Determine the linear velocity (feet/min) and its uncertainty.

2.4   The temperature of a certain liquid sample is measured to be 330 K with an uncertainty of ± 4 K.

(a) Convert the temperature and its uncertainty to Celsius.

(b) Convert the temperature and its uncertainty from Kelvin to Fahrenheit.

(c) Use your results from (a) and (b) to determine the relative uncertainties of this temperature in Kelvin, Celsius, and Fahrenheit. If your relative uncertainties differ from one another, explain how the dimensionless relative uncertainty can be affected by the units used to report results.

2.5 The pressure P exerted by a gas in a cylinder has been measured five times. A gage calibrated in psig was used. The results were $P_1 = 21.0$, $P_2 = 20.2$, $P_3 = 20.7$, $P_4 = 21.3$, and $P_5 = 20.5$.

   (a) From these measurements, estimate the Type A uncertainty.

   (b) The gage can be read to ± 0.1 psig and, during the readings, the room temperature was observed to fluctuate by ± 0.5 °C. Estimate the Type B uncertainty.

   (c) Now estimate the total uncertainty.

2.6 Two technicians from the analytic lab, Bert and Ernie, each report measurements for the density of a hydrocarbon mixture. Bert reports 0.835 g/cm^3 and Ernie reports 0.818 g/cm^3. Your experience is that Bert always measures too high, while Ernie always measures too low. Estimate the density and its Type B uncertainty.

2.7 Prove that the maximum of a Gaussian (2.4) coincides with its mean value (y_m). What is the meaning of the maximum?

2.8 The wall thickness, outside diameter (od), and inside diameter (id) of a pipe have each been measured five times, giving the values tabulated below. From these data, estimate the quantities (a)–(c):

Measure- ment #	od (inches)	id (inches)	Wall thickness (inches)
1	10.12	8.13	0.78
2	9.72	8.26	0.96
3	10.23	8.45	0.93
4	10.31	8.15	0.82
5	9.93	8.39	0.86

   (a) The outside diameter and its Type A uncertainty,

   (b) The inside diameter and its Type A uncertainty,

   (c) The wall thickness and its Type A uncertainty. Determine the wall thickness in two ways: (i) directly, using the data in the table, and (ii) indirectly, using your values for the id and od. Does one method give a more precise value for the wall thickness? If so, why?

2.9 Your production team is to prepare a binary mixture by combining N_1 moles of methanol with N_2 moles of water. The uncertainties in the measured mole numbers are estimated to be u_{N_1} for methanol and u_{N_2} for water.

(a) A team member claims that, since the mole fractions x_1 and x_2 must sum to unity, the uncertainties in the mole fractions must have the same magnitudes:

$$u_{x_1} = u_{x_2}$$

Do you believe this?

(b) Derive the relation between u_{x_1} and u_{x_2} for the situation in which the uncertainties in the mole numbers, u_{N_1} and u_{N_2}, are known.

(c) Regardless of whether the relation in (a) is obeyed, do you think that one component can usually be expected to have the smaller fractional uncertainty, u_{x_i}/N_i? For example, do you expect that the component having the smaller mole fraction will also have the smaller fractional uncertainty?

(d) Note from the relation found in (b) that both u_{x_1} and u_{x_2} depend inversely on the total number of moles N. This suggests that larger quantities of a mixture would have smaller uncertainties in the mole fractions than smaller quantities of the same mixture. Will this always be so?

2.10 The length of a brick wall has been measured by six people. Some used a yardstick, others used a six-foot flexible rule. The results were as follows:

Person #	1	2	3	4	5	6
Length (ft)	21.5	20.7	22.1	19.3	21.3	21.9

(a) Identify the possible outlier in these data.

(b) If you discarded the outlier, by what percentage would the mean change and by what percentage would the standard deviation change? Explain.

(c) Discuss what you would consider in trying to explain the outlier and whether you would, in fact, discard its value.

2.11 For a liquid having density ρ and viscosity μ flowing at linear velocity v through a pipe of diameter d, the Reynolds number Re is a dimensionless group defined by

$$\text{Re} = \frac{\rho v d}{\mu}$$

(a) Show how the uncertainty in Re would be computed from the uncertainties in the values measured for $v, \rho, d,$ and μ.

(b) In some situations, the linear velocity v is not measured directly; instead, the mass flow rate m is measured directly. Rewrite the above expression for Re in terms of the mass flow rate, then obtain an expression for the uncertainty in Re in terms of the uncertainties in $m, d,$ and μ.

(c) Comparing your result in (b) with that found in (a), would more accurate values for Re be obtained by measuring v or by measuring m? Or would the uncertainties in Re be about the same either way?

(d) Of the four independent quantities, do you expect that the uncertainty in one or two will often dominate the uncertainty in Re? To have a concrete example, let the fluid be liquid water. Then, would the uncertainties in some of the four be negligible when computing the uncertainty in Re? Justify your answer.

2.12 Do you expect that the standard deviation for one of several measurements will change as the number of measurements N is increased? What about the standard deviation of the mean?

2.13 Thelma and Louise each measure the length of a desktop. One uses a rigid rule calibrated in inches with the smallest division being 1/8 inch. The other uses a rigid ruler calibrated in centimeters, with the smallest division being 1 mm. Thelma reports the length as 59.9 ± 0.1 inches, while Louise reports 60.0 ± 0.04 inches. Was Thelma's ruler calibrated in inches or centimeters?

Exhibit C

The Prandtl number (Pr) is a dimensionless group of fluid properties often appearing in heat transfer correlations. Let c_p be the isobaric heat capacity, μ the shear viscosity, and k the thermal conductivity, then $Pr = (c_p \mu)/k$. The following plot shows experimental data for the temperature dependence of Pr for 19 refrigerants in the saturated vapor phase. Pr* is a reduced Prandtl number, $Pr^* = Pr/Pr_m$, where Pr_m is the minimum value for Pr in the saturated liquid state. T_c is the critical temperature. Line is an empirical correlation of the data.

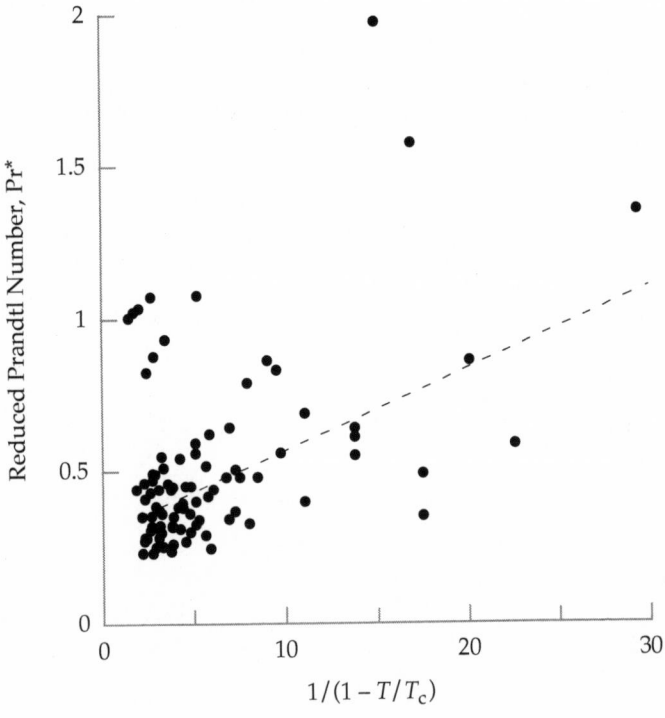

Data and line from M. W. Johnson and P. E. Liley, *Proceedings of the Seventh Symposium on Thermophysical Properties*, ASME, New York, 1977, p. 739.

3

Relating y to x

IN THE PREVIOUS CHAPTER WE DISCUSSED the measurement and repeated measurements of one value for a single variable. In this chapter we begin to consider experiments: activities by which we quantify relationships among variables. We restrict our attention to two variables, x and y. In the typical experiment, we set or control the value of the independent variable (x) and measure the corresponding value for the dependent variable (y). To capture the required behavior of y, we repeat this procedure over a range of x values.

The relation between x and y is usually quantified by a least-squares fit. But before we can perform such a fit, a functional relation between x and y must be known or assumed,

$$y = f(x) \tag{3.1}$$

The function f might be provided by the theoretical context under which the experiment is performed, but this is not always the case. Many times the theoretical context only suggests that x and y are related; then, a purpose of the experiment is to find the functional relation. Even when a functional form is provided by a theory, we might choose to test the theory by attempting to deduce the form for f from experimental data.

In any case, given data for x and y, we must identify the functional relation between the two. A systematic procedure for doing so is presented in § 3.1. Then we may fit our form to the data, as summarized in § 3.2. With the fit and the data, we may then perform additional analyses, including determination of residuals, as discussed in § 3.3.

3.1   Straightening x-y Plots

Given experimental data for y measured at known values of x, our problem is to identify the functional relation between the two. We invariably start by constructing an x-y plot. Our goal is to find a simple relation between x and y; that is, we seek a straight line. If our x-y plot is not straight, we try to straighten it by replotting, using other functions or other scales or both.

3.1.1 Searching for Straight Lines

Some functions readily linearize. One example is the exponential,

$$y = A e^{mx} \tag{3.2}$$

which linearizes on taking logarithms,

$$\ln y = \ln A + mx \tag{3.3}$$

That is, exponentials yield straight lines on semilog plots, as shown in Figure 3.1. Another example is the power law,

$$y = \alpha x^{\beta} \tag{3.4}$$

which also linearizes by taking logarithms,

$$\ln y = \ln \alpha + \beta \ln x \tag{3.5}$$

This means that power laws yield straight lines on log-log plots, as in Figure 3.2. Equations 3.3 and 3.5 are examples of the kinds of transformations we seek, though in general we can expect the required functions to be more complicated than exponentials or power laws. We need a systematic way to search for functions that linearize data. One way is to order simple functions of x in a progression, like this:

$$\cdots x^{-2} \cdots x^{-1} \cdots \ln x \cdots x \cdots x^2 \cdots x^3 \cdots$$

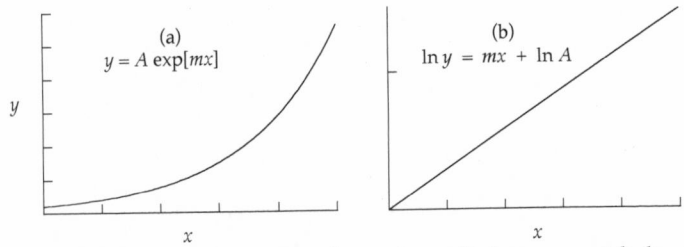

FIGURE 3.1 Exponential relations (a) between x and y become straight lines on semilog plots (b).

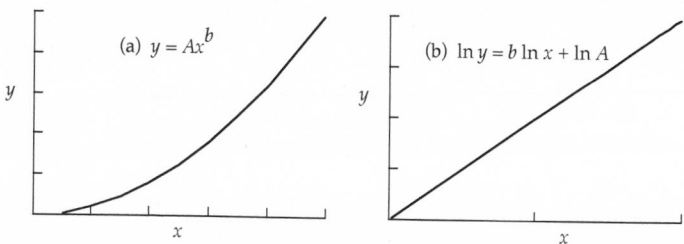

FIGURE 3.2 Power-law relations (a) between x and y become straight lines on log-log plots (b).

Then order functions of y in the same progression [1]:

$$\cdots y^{-2} \cdots y^{-1} \cdots \ln y \cdots y \cdots y^2 \cdots y^3 \cdots$$

Now we mimic the form of an x-y plot by arranging the progression in x horizontally and arranging that in y vertically, as in Table 3.1. We will use the functional progressions in Table 3.1 to guide our re-plotting of y vs x in our search for a linear relation.

Typically, we start from an x-y plot of the data. Then, depending on the signs of the first and second derivatives, we move along the progressions in Table 3.1 to find other functions of x and y that tend to straighten the curve. With two derivatives, each having one of two signs, we can identify four general possibilities.

TABLE 3.1 To make our search for straight lines systematic, we arrange functions of y in a vertical ladder (as on the ordinate) and arrange functions of x in a horizontal progression (as on the abscissa). Of course, fractional powers of x and y may also be tried in our attempts to linearize. After Tukey [1].

$$\vdots$$
$$y^3$$
$$y^2$$
$$y \qquad \cdots \quad x^{-2} \quad x^{-1} \quad \ln x \quad x \quad x^2 \quad x^3 \quad \cdots$$
$$\ln y$$
$$y^{-1}$$
$$y^{-2}$$
$$\vdots$$

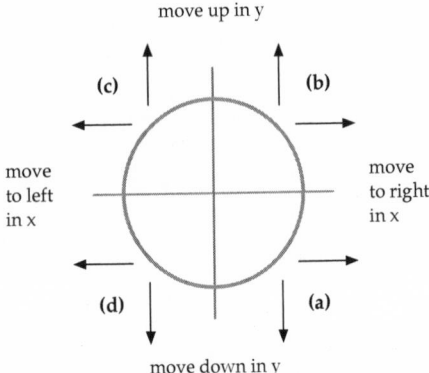

FIGURE 3.3   Each arc of this circle represents a possible shape of a monotonic, nonlinear curve obtained when y is plotted against x. To find new functions that tend to linearize the plot, we move along one of the progressions of functions shown in Table 3.1. The directions of movement are indicated by the arrows. After Tukey [1].

(a) If dy/dx and d^2y/dx^2 are both positive, then move either to the right along the progression in x or down along the ladder in y. See Figure 3.3(a).

(b) If both derivatives are negative, then move either to the right along the progression in x or up along the ladder in y. See Figure 3.3(b).

(c) If $dy/dx > 0$ while $d^2y/dx^2 < 0$, then move either to the left along the progression in x or up along the ladder in y. See Figure 3.3(c).

(d) If $dy/dx < 0$ while $d^2y/dx^2 > 0$, then move either to the left along the progression in x or down along the ladder in y. See Figure 3.3(d).

Usually we must repeatedly apply the transformations in Table 3.1 to find a (nearly) straight line. When this is done, it may help to let the current $f(y)$ become y on Table 3.1 and let the current $f(x)$ become x.

Note that we may change the function of x or that of y or both. Some data will be more sensitive to changes in x, while other data will be more sensitive to changes in y. Often, the functions of both x and y must be changed to achieve a (nearly) straight line.

3.1.2 A Measure of Linearity

We can quantify the linearity of x-y plots, such as those in Figure 3.1, by computing a linear correlation coefficient, r, which is defined as

$$r = \frac{\sum_i^N (x_i - x_m)(y_i - y_m)}{\sqrt{\left(\sum_i^N (x_i - x_m)^2\right)}\sqrt{\left(\sum_i^N (y_i - y_m)^2\right)}} \quad (3.6)$$

Here x_m is the mean of the N values of the independent variable x_i and y_m is the mean of the N measured values of the dependent variable y_i. Values of r lie on $[-1, 1]$. A value of $r = 1$ implies a perfectly straight line with positive slope; $r = -1$ indicates a straight line with negative slope; $r = 0$ signals a straight horizontal line, which means that the value of y is not correlated with x.

Data having values of $|r|$ close to unity are more linear than data with $|r|$ close to zero. However, many kinds of nonlinearities are possible and the coefficient r cannot distinguish among them. For example, different nonlinearities can produce the same value for r, as in Figure 3.4. So at best, the linear correlation coefficient is only a crude measure of linearity.

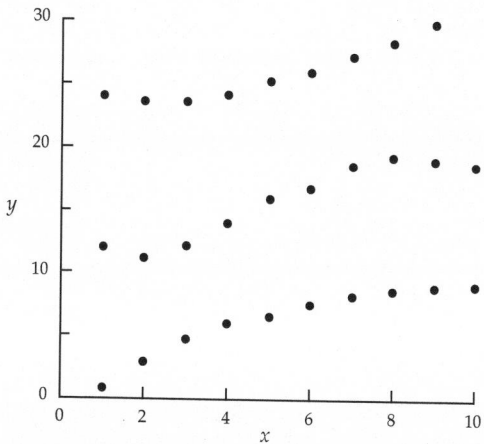

Figure 3.4 The linear correlation coefficient only crudely measures nonlinearity. Here, for example, are three sets of data that differ in their nonlinearities, although all three sets have the same value of $r = 0.950$.

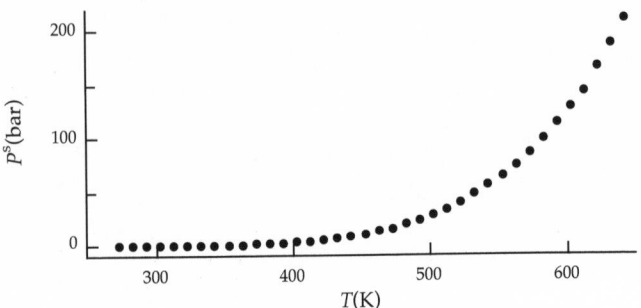

FIGURE 3.5   Vapor pressures of water from the triple point (273.16 K) to the critical point (647.10 K). $r = 0.85$. Data from NIST Webbook [2].

3.1.3   Example: Vapor Pressures of Water

To illustrate the use of the ideas from § 3.1.1 and 3.1.2, consider the vapor pressures P^s of water, plotted against absolute temperature T in Figure 3.5. This is a highly nonlinear plot; in fact, the value of the linear correlation coefficient is $r = 0.85$. To straighten the plot, we compare the shape in Figure 3.5 with each quadrant of the diagram in Figure 3.3. The dotted line has both dP^s/dT and $d^2P^s/dT^2 > 0$; that is, it has the shape of the curve in the fourth quadrant of Figure 3.3. This means we could straighten the curve by moving down the vertical ladder in y or by moving to the right on the horizontal progression in x. We choose to move down the ladder from y to $\ln y$, so we replot the data in the form $\ln(P^s)$ vs T.

This new plot is shown in Figure 3.6. Note in the figure that we obtain $\ln(P^s)$ by using a log scale on the ordinate. The curve in Figure 3.6 has a linear correlation coefficient of $r = 0.97$, so it is significantly

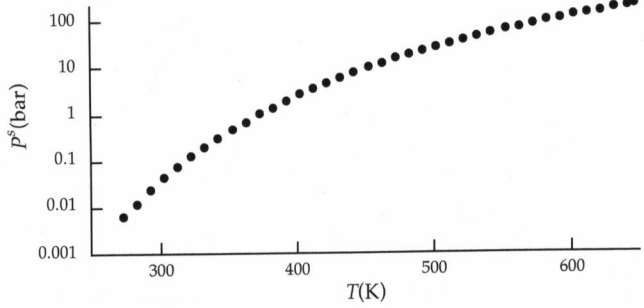

FIGURE 3.6   Vapor pressures from Figure 3.5, replotted on semilog axes in an attempt to make the curve linear. $r = 0.97$.

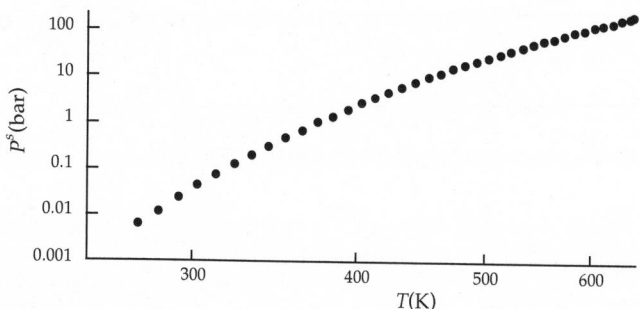

FIGURE 3.7 Vapor pressures from Figure 3.5, plotted on log-log axes; the curve is still not straight. $r = 0.990$

straighter than the curve in Figure 3.5, but it still retains some curvature. However, the curvature in Figure 3.6 is opposite to that in Figure 3.5: we have over-corrected in our attempt to straighten the curve.

The shape of our new curve in Figure 3.6 corresponds to that in the second quadrant of Figure 3.3. To straighten it, we must (according to Figure 3.3) move up the ladder of expressions in y or move to the left along the progression of expressions in x. Since we don't want to undo what we have accomplished by moving in y, let's now change the function for x. So we move from x to $\ln x$ on the progression in Table 3.1: we plot $\ln(P^s)$ vs $\ln(T)$, as in Figure 3.7.

The curve in Figure 3.7 has $r = 0.990$, so it is straighter than that in Figure 3.6, but some curvature remains. Therefore, we move farther to the left on the progression in x, from $\ln x$ to $1/x$. For example, if we plot $\ln(P^s)$ vs $1/T$, as in Figure 3.8, we obtain a nearly straight line. Indeed, the curve in Figure 3.8 has $r = 0.9997$.

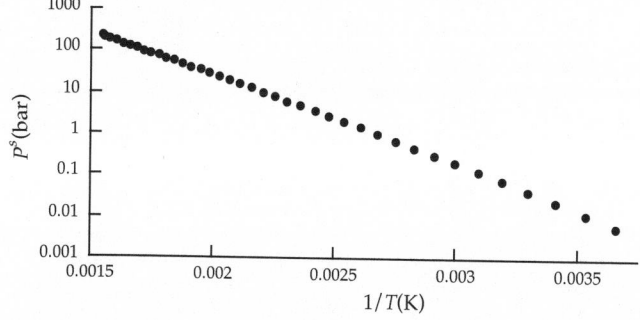

FIGURE 3.8 Vapor pressures of water plotted to obtain a nearly straight line. $r = 0.9997$. Data from Figure 3.4.

The plot in Figure 3.8 is consistent with a fundamental thermodynamic relation, the Clausius-Clapeyron equation, which suggests that straight lines should result when the log of the vapor pressure is plotted against reciprocal absolute temperature. Our purely mathematical manipulations of the original data have produced a result that is consistent with a theoretical context provided by thermodynamics.

3.2   Simple Least-Squares Fits

Once we have found functions of x and y that are related (more or less) linearly, we need to quantify the relation; that is, we need to find the slope and intercept of a representative line. But which line shall we use? Although we have found a linear representation of our data, using the procedure in § 3.1, some scatter remains because of the errors discussed in § 2.1. Further, the linearization is probably only approximate, not exact. So any of an essentially infinite number of straight lines might be chosen to represent the data. We need criteria that can be imposed systematically and that produce a unique straight line from our data.

3.2.1   The Normal Equations

Consider Figure 3.9, which shows five measured points and an arbitrary straight line that might represent the measurements. We assume the usual experimental situation in which we have controlled the values of x and measured corresponding values for y. Hence, the uncertainties all reside in the y-values. We have estimated the quality of the measurements, as discussed in Chapter 2, so we have an uncertainty u_i assigned to each measured y_i; we use those uncertainties to weight the data.

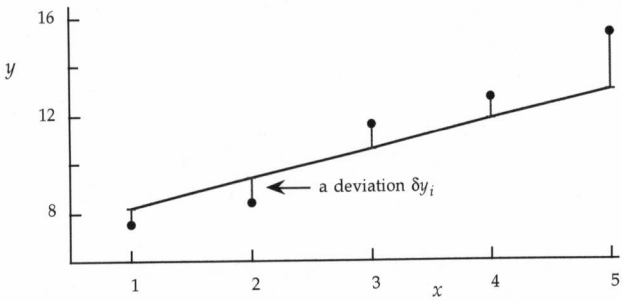

FIGURE 3.9   Example of five points $y(x)$ that might be correlated by a straight line. One possible straight line is shown.

Define the deviation δy_i to be the vertical distance between a measured point y_i and the corresponding y-value on the straight line (One such deviation is labeled on Figure 3.9.),

$$\delta y_i = y_i - (mx_i + b) \tag{3.7}$$

Here m is the slope and b is the intercept of the line. Of the many lines we might use, we choose the one that minimizes the sum of the squares of the weighted deviations. That is, we have a minimization problem with respect to the slope and intercept:

$$\underset{\{m,b\}}{\text{Min}} = \sum_i^N \left(\frac{\delta y_i}{u_i}\right)^2 \tag{3.8}$$

By placing the uncertainties in the denominator, we weight measurements with small uncertainties more than the others. The minimization problem (3.8) is solved by forming the derivatives with respect to m and b, then setting each to zero:

$$\frac{\partial}{\partial m}\left[\sum_i^N \left(\frac{\delta y_i}{u_i}\right)^2\right] = 0 \tag{3.9}$$

$$\frac{\partial}{\partial b}\left[\sum_i^N \left(\frac{\delta y_i}{u_i}\right)^2\right] = 0 \tag{3.10}$$

These are two equations that can be solved for two unknowns: m and b. In fact, equations (3.9) and (3.10) are linear in m and b, so they can be solved using linear algebra. The results are

$$m = \frac{SS_{xy} - S_x S_y}{\Delta} \tag{3.11}$$

$$b = \frac{S_{xx}S_y - S_x S_{xy}}{\Delta} \tag{3.12}$$

$$\Delta = SS_{xx} - (S_x)^2 \tag{3.13}$$

Equations (3.11)–(3.13) are called the *normal least-squares equations* [3]. In these equations, S_{xx}, etc. represent the following five sums [3]:

$$S = \sum_i^N \frac{1}{u_i^2} \qquad S_x = \sum_i^N \frac{x_i}{u_i^2} \qquad S_y = \sum_i^N \frac{y_i}{u_i^2} \qquad (3.14)$$

$$S_{xx} = \sum_i^N \frac{x_i^2}{u_i} \qquad S_{xy} = \sum_i^N \frac{x_i y_i}{u_i^2} \qquad (3.15)$$

Here, N is the number of measured points being fitted. For the five points shown in Figure 3.9, all weighted equally (so we set all $u_i = 1$), the normal equations give the least-squares line as

$$y = 1.99x + 5.13 \qquad (3.16)$$

This line is shown in Figure 3.10. Substituting into (3.7) each y-value from Figure 3.10 together with the corresponding y-value from the line (3.16), we find the sum-of-squares of deviations to be

$$\sum_{i=1}^5 (\delta y_i)^2 = 2.1475 \qquad (3.17)$$

For the five equally-weighted points in Figure 3.10, the least-squares procedure assures us that no other straight line has a smaller sum-of-squares of deviations than the value in (3.17).

If the data in Figure 3.10 were not all equally reliable, then the weights would not all be the same and the normal equations would yield some other straight line. For example, if the uncertainties increase with x and y, as in Figure 3.11, then the resulting least-squares line becomes

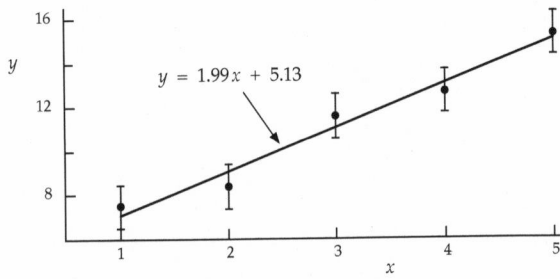

FIGURE 3.10 Same five points as in Figure 3.9, with least-squares line (3.16) computed from normal equations (3.11)–(3.15) using equal weights (all $u_i = 1$). Error bars are expanded uncertainties at the 68% confidence level.

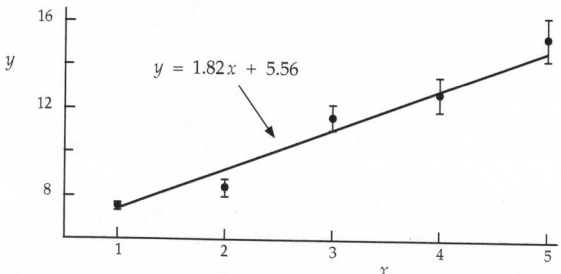

FIGURE 3.11 For the same data as in Figure 3.10, but with uncertainties that increase with x, the least-squares line (3.18) differs from the one in Figure 3.10. Error bars are expanded uncertainties at the 68% confidence level.

$$y = 1.82x + 5.56 \tag{3.18}$$

Figure 3.11 shows that this line lies near the most reliable point (at $x = 1.0$) and not so near the least reliable point (at $x = 5.0$).

In contrast, if the uncertainties decrease with increasing x and y, as in Figure 3.12, then the normal equations give

$$y = 2.09x + 4.79 \tag{3.19}$$

The line in Figure 3.12 lies close to the reliable point at $x = 5.0$, and not so close to the less reliable point at $x = 1.0$. Figures 3.10–3.12 show that we can bias least-squares lines by including uncertainties in the fits. In this way we can force fits to reproduce those portions of the data that are most reliable.

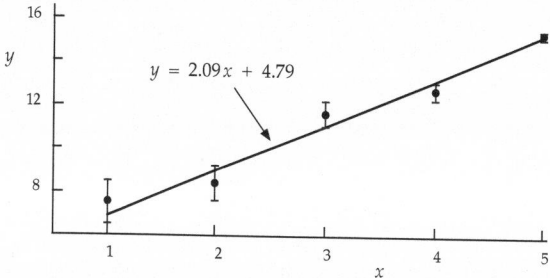

FIGURE 3.12 For the same data as in Figures 3.10 and 3.11, but with uncertainties that decrease with increasing x, the least-squares line (3.19) differs from both that in Figure 3.9 and that in Figure 3.11. Error bars are expanded uncertainties at the 68% confidence level.

When all points are weighted equally, the least-squares procedure tends to place more emphasis on points at the ends of the range. But in many experiments, extreme values are measured less reliably than other values. Therefore, it may be important to give special attention to extreme values and weight their deviations accordingly.

Similar comments apply to outliers. Outliers have deviations that are anomalously large, and since least squares is trying to minimize the sum of squares of deviations, the least-squares calculation tends to unduly emphasize the importance of outliers. So if you have a probable outlier, but can't reach a decision to ignore it completely, then consider weighting it less than other points.

3.2.2 Comments on Least Squares

One of the normal equations (3.9) forces the sum of the (weighted) deviations to be exactly zero (within round-off error),

$$\sum_i^N \frac{\delta y_i}{u_i^2} = 0 \qquad (3.20)$$

This provides a convenient check on whether the least-squares calculations have been done correctly. Note that (3.20) is a necessary but not sufficient test for correctness.

The least-squares line has a smaller sum of squares of deviations than any other line that could be drawn through the given N points. However, if you perform more measurements, increasing the number of points N, then your calculated line may not be the best representation of all the data: the least-squares solution should be recomputed using all the data.

Once the fit has been performed, then the propagation of uncertainty procedure (§ 2.5) can be applied to find the uncertainties in the fitted slope and intercept [3]. The results of that procedure give the uncertainty in the slope as

$$u_m = \sqrt{\frac{S}{\Delta}} \qquad (3.21)$$

and the uncertainty in the intercept as

$$u_b = \sqrt{\frac{S_{xx}}{\Delta}} \qquad (3.22)$$

From these we can show that both uncertainties are inversely proportional to the root of the number of measurements N,

$$u_m \propto \frac{1}{\sqrt{N}} \quad \text{and} \quad u_b \propto \frac{1}{\sqrt{N}} \tag{3.23}$$

This is an example of the *law of large numbers*: if we want to decrease the uncertainties in the slope and intercept by a factor of two, then we must increase the number of measurements by a factor of four.

In reporting the least squares line, many people also report the value of the linear correlation coefficient r, defined in (3.6). But the definition (3.6) for r contains nothing about the least-squares line; hence, r can be computed before the fit is done, as suggested in § 3.1.2. Then r can be used to help judge whether we have adequately linearized the data. For the five points in Figure 3.9, $r = 0.986$.

3.2.3 Generalized Least Squares

In § 3.2.1 we presented the normal least-squares equations for fitting a straight line to x-y data. But the requirement for applying linear least squares is that the fitting function be linear in the unknown parameters; it need not be linear in x. This means that the approach extends to any polynomial representation of $y(x)$, as shown in Appendix B.

However, a polynomial is not always the best way to represent nonlinear data. In such cases, a linear least-squares fit can still be done if a suitable transformation can be found that leaves $y(x)$ in a linear form,

$$f(y_i) = m g(x_i) + b \tag{3.24}$$

To perform the fit, we still use the normal equations (3.11)–(3.15), but we replace y with $f(y)$ and x with $g(x)$,

$$y_i \rightarrow f(y_i) \quad \text{and} \quad x_i \rightarrow g(x_i)$$

For example, if we find that a semilog plot linearizes the data, then the functional form is exponential,

$$y = A \exp[mx] \tag{3.25}$$

We linearize by taking the log of both sides,

$$\ln y = \ln A + mx \tag{3.26}$$

and we perform the least squares fit on $\ln(y)$ vs x. The results from that fit are the slope m and the intercept $\ln(A)$.

To illustrate, we fit the vapor pressure data in Figure 3.8. The figure shows that the data are approximately linearized by

$$\ln P^s = \frac{m}{T} + b \tag{3.27}$$

This implies we are to perform a least-squares fit using $y = \ln(P^s)$ and $x = 1/T$. For P^s in bar and T in Kelvin, the normal equations (3.11)-(3.15) yield the line shown in Figure 3.13,

$$\ln P^s = 13.03 - \frac{4891}{T} \tag{3.28}$$

At $100°\text{C} = 373.15$ K, the fitted correlation (3.28) gives a vapor pressure of 0.926 bar. But at this temperature, we know that the vapor pressure of water is 1.013 bar, so the correlation is in error by almost 9%. Note that Figure 3.13 is, in a real sense, deceptive: the logarithmic and reciprocal scales fail to give us a sense that the line and points disagree by 9% near $1/T = 0.0027$.

A 9% error is understandable, considering that the fit was done over a temperature range of almost 375 K degrees. Whether or not a 9% error is tolerable depends on the use to be made of the fit. If it is unacceptable, then we must either find a nonlinear representation of the data or refit (3.27) to a more restricted range of temperatures.

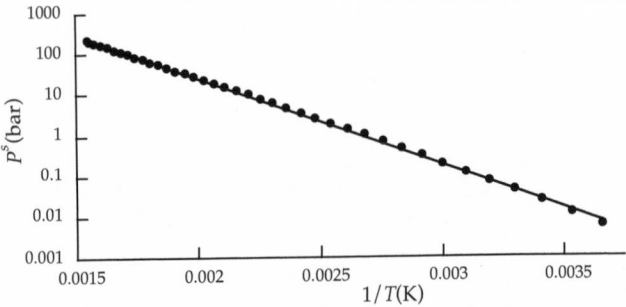

FIGURE 3.13 Vapor pressures for water. Points are from NIST Webbook [2] and are same as in Figure 3.8. Line is the least-squares fit (3.28).

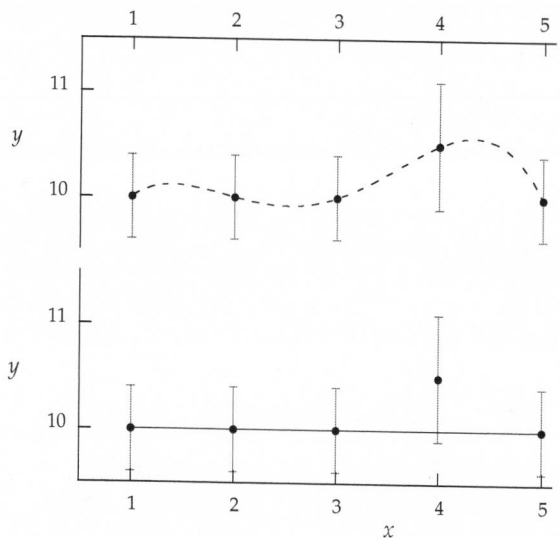

FIGURE 3.14 (top) Through five points we can exactly fit a fourth-order polynomial (broken line at top). However, for these data, a straight line (bottom) is sufficient to correlate the measurements within their uncertainties. Error bars are expanded uncertainties at the 68% confidence level.

Since least-squares fits are relatively easy to perform, it is tempting to apply them indiscriminately; that is, it is often easy to simply choose a nonlinear function that appears to correlate the data, then force the fit by solving the least-squares problem. Such activities are common, but they are not sound engineering. Engineering requires judgment, and in correlating data, the judgment occurs in choosing the proper function and in assigning weights to the data. These decisions must be made before a least-squares calculation can be done.

When choosing a function, we should be guided, not only by the pattern appearing on an x-y plot, but also by the uncertainties assigned to the measurements. For example, consider the five points plotted in the top panel of Figure 3.14. With five points we can compute values for five unknowns; hence, we can force a fourth-order polynomial to pass exactly through those five points. That curve is also shown at the top of Figure 3.14. But note that the polynomial must oscillate if it is to exactly reproduce the original data. (Such oscillations are common to high-order polynomial fits.) Unless we have information that supports the presence of such oscillations, we can only justify the least complex function that reproduces the data within their uncertainties. For the data and uncertainties in Figure 3.14, that function is a straight line, as at the bottom of the figure.

3.3   Residuals

Given a set of x-y data, let's assume we have linearized the data (as in §3.1) and have fit a least-squares line to the linear form (as in §3.2). Although we now have an approximate representation of the data, that representation will not be exact: there are errors in the measurements, and the relation between x and y may be more complicated than our (relatively) simple straight line. To explore these possibilities further, we need to amplify the remaining nonlinearities. This is often done by computing residuals: either absolute residuals (§3.3.1) or fractional residuals (§3.3.2) or both.

3.3.1   Absolute Residuals

An absolute residual is the difference between a measured y-value and the corresponding fitted y-value:

$$\Delta y_i = y_{i,\text{mea}} - y_{i,\text{fit}} \tag{3.29}$$

To illustrate, consider Figure 3.15, which shows product concentration at discrete times during a chemical reaction. Using the procedure from

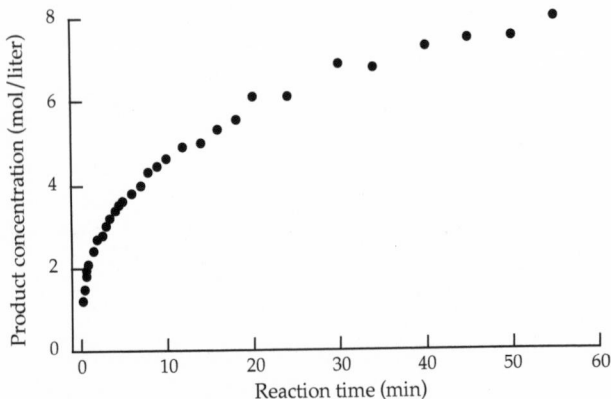

FIGURE 3.15   Experimental data for product concentration c at discrete times t during a chemical reaction at $35°$C.

§ 3.1.1, we find that these data can be linearized by plotting on log-log axes, as in Figure 3.16. Assuming equally weighted points and fitting a power law, we obtain

$$\frac{c}{\text{mol/l}} = 2.096 \left(\frac{t}{\text{min}} \right)^{0.3375} \tag{3.30}$$

Units are included in (3.30) to make the lhs and rhs both dimensionless; consequently, the coefficient (2.096) and the exponent (0.3375) are dimensionless. If the reaction time is allowed to increase from t_1 to t_2, then the corresponding change in concentration is given by (3.30) as

$$\frac{c_2}{c_1} = \left(\frac{t_2}{t_1} \right)^{0.3375} \tag{3.31}$$

which is dimensionless and independent of the units chosen for c and t. Equation (3.31) shows that if the reaction time is allowed to increase by a factor of ten $(t_2/t_1 = 10)$, then the product concentration will approximately double: $c_2/c_1 = 10^{0.3375} \approx 2.2$.

However, we may question whether all the data in Figure 3.16 are equally reliable; that is, are we justified in weighting all points equally? The log scales in Figure 3.16 may obscure discrepancies between measurements and the fitted line; therefore, we can appeal to residuals to

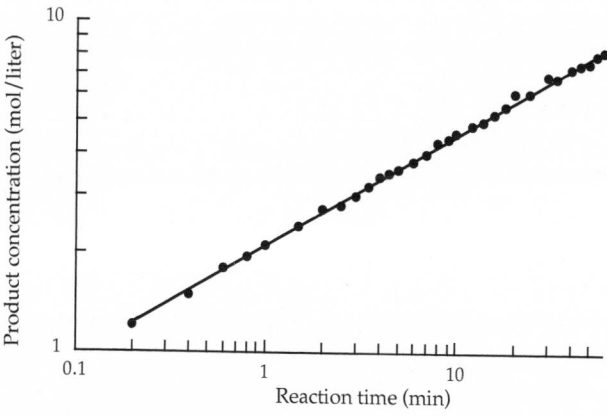

FIGURE 3.16 Same data as in Figure 3.15 but on log-log axes; line is the least-squares fit (3.30).

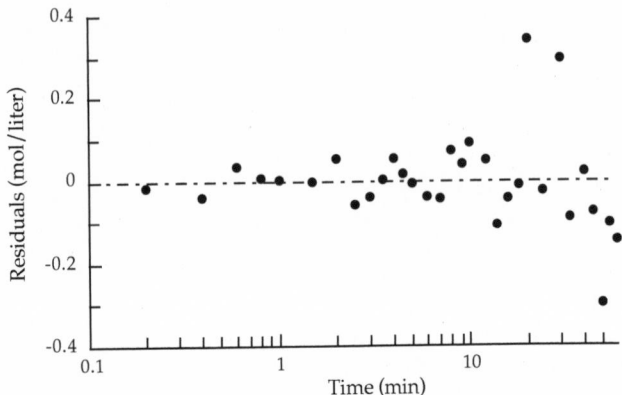

FIGURE 3.17 Absolute residuals from the points and line in Figure 3.16.

amplify any discrepancies. Using the data from Figure 3.16 and the line from (3.30), we form the absolute residuals, which are plotted in Figure 3.17. In the figure, a positive (negative) residual indicates a measured point that is greater than (less than) the corresponding fitted point; see (3.29). Note that, at long times, the residuals in Figure 3.17 show some large discrepancies that are not apparent in Figure 3.16.

From a plot of residuals, such as in Figure 3.17, we seek to answer these kind of questions [1]:

(a) What is the range of the residuals?

(b) How are the residuals distributed about zero?

(c) How do the residuals change with x? Are there any patterns?

(d) Are the magnitudes of some residuals unusual compared to the magnitudes of others?

Answers to these can help us judge the quality of the data and the reliability of the linearizing function; such possibilities are summarized in Table 3.2. If we apply the guides in Table 3.2 to the above questions for the residuals in Figure 3.17, we obtain these observations:

(a) The residuals in the figure range from -0.30 mol/l to $+0.34$ mol/l, so the magnitude of the range is 0.64 mol/l. The mean concentration is 4.5 mol/l, so the range is about 14% of the mean. That is, the absolute value of the range is modest, but its value relative to the mean seems somewhat large.

TABLE 3.2 Possible ways to interpret the behavior of residuals when they are plotted against an independent variable x, as in Figure 3.17. We caution that these are not definitive interpretations, but merely suggestions to help start your thinking.

Feature	Behavior	Possible Meaning
Range	(a) range wide compared to mean y-value	(a) large errors in data or poor choice for linearizing function
	(b) range narrow compared to mean y-value	(b) small statistical error, but systematic error may be present
Distribution	(a) random scatter of residuals about zero	(a) error probably dominated by statistical errors, but constant systematic error may be present
	(b) nonuniform scatter about zero	(b) bias in fit due to unequal weights or error in solving least-squares problem
	(c) systematic variation	(c) systematic error in data or some nonlinearity not captured by linearizing function
	(d) magnitudes of residuals small except at large x or small x or both	(d) large systematic error at extremes or linearizing function fails at extremes
Patterns	(a) magnitudes of residuals increase with y	(a) possible constant relative error; check fractional residuals
	(b) residuals oscillate about zero	(b) periodic systematic error or poor linearizing function
Unusual magnitudes	(a) some closely grouped residuals have magnitudes that differ from others	(a) some measurements may be anomalous over a small range of x values
	(b) a few residuals have relatively large magnitudes	(b) possible outliers

(b) Of the 31 points in Figure 3.17, the distribution about zero is roughly even: 14 are positive and 17 are negative. However, their variation does not appear random; rather, the residuals appear to oscillate systematically about zero.

(c) The magnitudes of the residuals seem to be slowly increasing with time.

(d) Three points (those at 20, 30, and 50 minutes) appear to have much larger magnitudes than the other 28 points. Indeed, if those three points are ignored, then the remaining 28 residuals lie between -0.15 mol/l and $+0.10$ mol/l, which is less than 6% of the mean concentration.

These observations suggest that the measurements contain some systematic error that increases with time. They also suggest that three of the 31 points are possible outliers, to which we would apply Chauvenet's criterion from § 2.3.4.

3.3.2   Fractional Residuals

In Figure 3.17 the magnitudes of the residuals are increasing with time; hence, by Figure 3.16, they are increasing with concentration. This behavior is common; that is, the magnitudes of residuals often increase as the magnitude of the dependent variable increases. This may be caused by a constant relative error in the data. To test for this, you should compute fractional residuals,

$$\frac{\Delta y_i}{y_i} = \frac{y_{i,\text{mea}} - y_{i,\text{fit}}}{y_{i,\text{mea}}} \tag{3.32}$$

For the data in Figure 3.16, the fractional residuals are plotted in Figure 3.18. Compared with the absolute residuals in Figure 3.17, those in Figure 3.18 show a more uniform distribution about zero, implying a constant relative error. Further, Figure 3.18 suggests that the measurements at $t = 20$ and 30 minutes are outliers, but it is less certain whether that at 50 minutes is also one.

     Besides the uses suggested in Table 3.2, plots of residuals can sometimes clarify behavior and trends that might be overlooked or misinterpreted from simple x-y plots. To illustrate, consider the x-y plot in Figure 3.19; in the figure, the line is an unweighted least-squares fit to the points. Comparing the points with the line, we see more scatter and larger deviations in the points at large x relative to those at small x

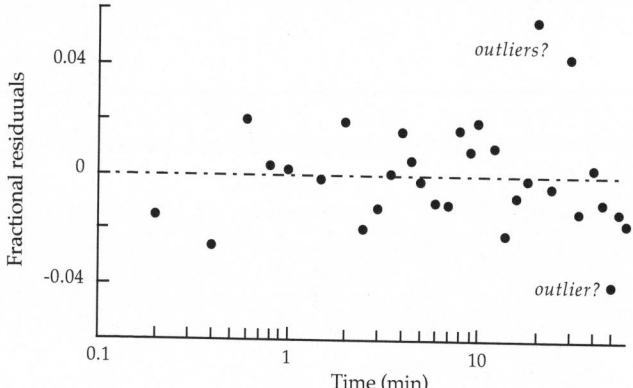

FIGURE 3.18 Fractional residuals (3.32) from line and points in Figure 3.15.

values. However, the human eye has trouble judging vertical distances between points and a nearly vertical line ($x < 2$ in the figure). To compensate for this, we plot in Figure 3.20 the absolute residuals between the points and the line. Figure 3.20 clearly shows that the points at small x are, in fact, farther from the line than are the points at large x.

However, instructive as it is, Figure 3.20 does not reveal everything hidden in the original data. To probe further, we form the fractional residuals from the points and line in Figure 3.19; moreover, we ignore the signs of those residuals and focus on their magnitudes. That is, in Figure 3.21 we plot the absolute values of the fractional residuals,

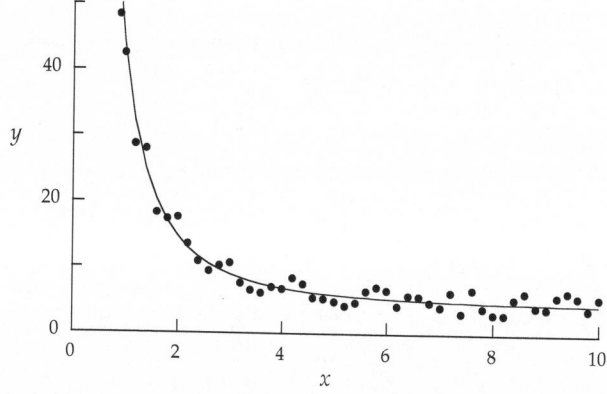

FIGURE 3.19 Simulated x-y data with least-squares fit (line). Points at small x ($x < 2$) appear closer to the line than points at large x.

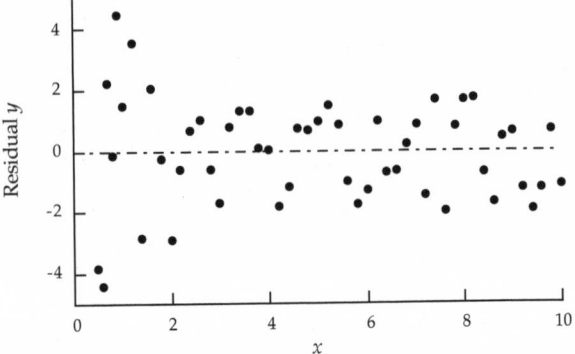

FIGURE 3.20 Absolute residuals from Figure 3.19 showing that points at small x, in fact, deviate more from the line than points at large x.

$$\frac{|\Delta y_i|}{y_i} = \frac{|y_{i,\text{mea}} - y_{i,\text{fit}}|}{y_{i,\text{mea}}} \tag{3.33}$$

Figure 3.21 suggests that the magnitudes of the fractional residuals are increasing slowly with time, although, with four exceptions, the magnitudes remain below 0.3. The three points near $x = 8$ are probably outliers, but it is less certain whether the fourth near $x = 5$ is also one. The lesson from Figures 3.19–3.21 is that simple x-y plots may fail to reveal important features, and we must do some creative exploration, by replotting the data in various forms, so as to amplify hidden features and make them visible.

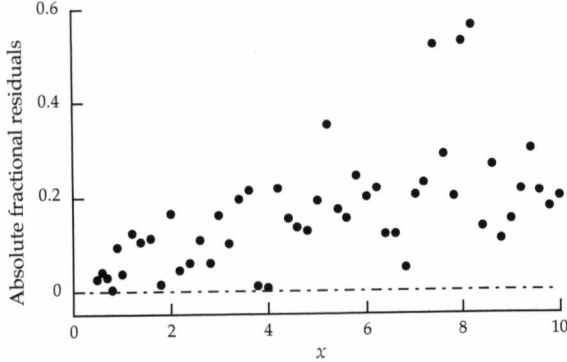

FIGURE 3.21 Absolute values of fractional residuals from Figure 3.18 reveal three probable outliers near $x = 8$ and possibly a fourth near $x = 5$.

Literature Cited

[1] J. W. Tukey, *Exploratory Data Analysis*, Addison-Wesley, Reading, MA, 1977.

[2] W. G. Mallard and P. J. Linstrom, eds., *Chemistry Webbook, NIST Standard Reference Database Number 69*, Feb. 2000, National Institute of Standards and Technology, Gaithersburg, MD. Website at (http://webbook.nist.gov/chemistry/).

[3] W. H. Press, B. P. Flannery, S. A. Teukolsky, and W. T. Vetterling, *Numerical Recipes*, Cambridge University Press, Cambridge and New York, 1986.

Exercises

3.1 Rewrite each of the following to obtain a linear relation between some function of x and some function of y. Identify the slope and intercept in your linearization.

(a) $y = a \exp[x - c]$

(b) $y = ax^{-b}$

(c) $y = a\sqrt{x} - c$

(d) $y = \dfrac{a}{x + c}$

3.2 Find the appropriate functions of x and y that plot to give a nearly straight line from the following data:

x	1	2	3	4	5	6
y	13.5	6.4	3.7	2.3	1.45	1.05

3.3 For your linearized form of the data in Problem 3.2, (a) find the least-squares line, assuming equally weighted points, and (b) determine the uncertainties in your computed slope and intercept.

3.4 From the data in Problem 3.2 and the least-squares line in Problem 3.3, compute and plot the absolute residuals. Use your plot to answer the four questions about residuals posed below Figure 3.17.

3.5 Rewrite the sums (3.14)–(3.15) for the special case in which all N points are weighted equally.

3.6 Find the appropriate functions of x and y that plot to give a nearly straight line from the following data:

x	100	150	200	300	400	500	600
y	−160	−71.5	−35.2	−4.2	9.0	16.9	21.3

3.7 Prove that, for N equally weighted points, the least-squares line given by the normal equations (3.11)–(3.15) forces the sum of the deviations to zero; that is, we must have

$$\sum_i^N \delta y_i = 0$$

where δy_i is given by (3.7). Now consider these four (x, y) pairs: $(1, 2.1)$, $(2, 3.8)$, $(3, 5.9)$, $(4, 8.1)$. For these four points find two different straight lines that each give the sum of squares of the deviations equal to zero. Conclusion?

3.8 A colleague says that, since for equally weighted points the sum of the deviations must be zero (see Problem 3.7), we can expect that the number of points having positive deviations from the least-squares line will be the same as the number having negative deviations, when the total number N is even. When the total number is odd, the number of positive deviations will differ only by unity from the number of negative deviations; Figures 3.9, 3.10, and 3.11 provide examples.

(a) Do you believe it?

(b) Consider data for which, over a restricted range of the independent variable x, the number of residuals having one sign differs significantly from the number having the other sign. An example occurs in Figure 3.17 for $t \geq 20$ min. What might this suggest about the measurements? What might it suggest about the fit?

3.9 Prove that when a least-squares straight line passes through the origin (so $b = 0$), then the normal equations in (3.11)–(3.15) reduce to $m = S_y/S_x$.

3.10 Perform a least-squares fit of the following data to the form:

$$y = ax^2 + b$$

Then estimate the uncertainties in your values obtained for the slope and intercept.

x	1	1.5	2	2.5	3	4
y	0.7	5.1	11.0	17.9	26.0	52.5

3.11 The following questions apply to these data:

x	1	2	3	4	5	6
y	2.2	4.0	5.8	7.0	8.1	9.4

(a) Perform a least-squares fit of the data to $y = ax^{0.75}$ and then report the sum of squares of deviations between the data and your fitted line.

(b) Now fit the data to ($\ln y = A \ln x + B$) and report the sum of squares of deviations between the data and this line.

(c) Explain any difference between the sums of squares of deviations found in (a) and (b).

(d) Compute and plot the absolute residuals formed from the data and the fitted line found in (a). Use Table 3.2 to access the quality of the data and the reliability of the fit.

(e) Compute and plot the fractional residuals formed from the data and the fitted line found in (b). Use Table 3.2 to access the quality of the data and the reliability of the fit.

Exhibit D

Test for correlation between the Dow-Jones industrial stock average and the distance from comet Halley to sun on 1 July of each year from 1947 (aphelion) to 1986 (perihelion). Distances in astronomical units (au), with 1 au = mean distance between earth and sun. On this plot, the data have a linear correlation coefficient of $r = 0.94$. Line is a least-squares fit.

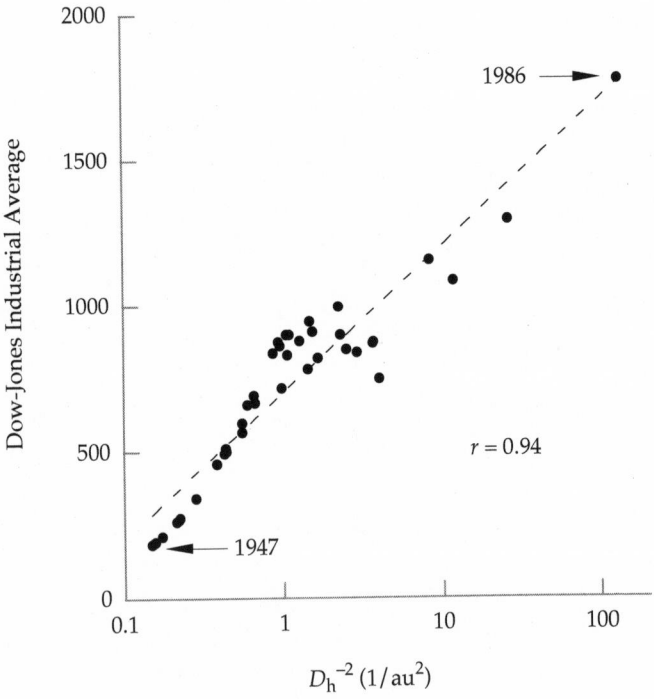

Values of the Dow-Jones average (on ordinate) taken from *Dow Data* at the Dow-Jones Index website (http://www.djindexes.com/). Distances (on abscissa) computed using the *Ephemeris Generator* at JPL Horizons On-Line Solar System Data and Ephemeris Computation Service (http://www.ssd.jpl.nasa.gov/).

4

Are x and y Correlated?

IN THE STANDARD EXPERIMENT WE study the relation between x and y by manipulating x and measuring how y responds. For example, we may change the flow rate to a heat exchanger and measure how a temperature responds. But there are situations in which this approach fails to quantify the relation between x and y. Two situations are common:

1. Scatter in the measured data may be so large that we cannot, with confidence, establish the relation between x and y. Sometimes the scatter may be such that we doubt whether x and y are even related at all.

2. We cannot manipulate or control either x or y; we can only make measurements. In these situations we can only measure both x and y and then try to establish a relation by correlation. For example, we might need to determine whether pollution in a river is being caused by effluents from our manufacturing plant.

In this chapter we focus on the problem of deciding whether x and y are correlated: Is there a relation between the two? This is a qualitative problem of some subtlety. For example, even when we can establish that a correlation exists, we have not necessarily established that a cause-and-effect relation exists between x and y. Many people try to attack this problem using the linear correlation coefficient r (§ 4.1), but we argue that there is a better way (§ 4.2).

4.1   Linear Correlation Coefficients

If x and y are correlated, then any of many possible functional forms might describe that correlation. Of the possibilities, we consider the linear form. If the relation is nonlinear, it might be linearized using the procedures in § 3.1; so, the material in this section is more general than it may appear.

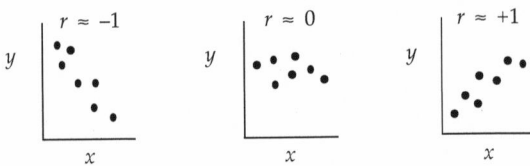

FIGURE 4.1   Schematic plots showing negative linear correlation (left), no correlation (center), and positive linear correlation (right).

In § 3.1.2 we introduced the linear correlation coefficient r as a tool for quantifying linearity. Recall that the quantity r is defined by

$$r = \frac{\sum_i^N (x_i - x_m)(y_i - y_m)}{\sqrt{\left(\sum_i^N (x_i - x_m)^2\right)}\sqrt{\left(\sum_i^N (y_i - y_m)^2\right)}} \qquad (4.1)$$

Here (x_i, y_i) are the N measured pairs of data; x_m is the mean of the xs and y_m is the mean of the ys. Values of r lie between -1 and $+1$. Values near -1 imply a negative linear correlation, values near $+1$ signal a positive linear correlation, and values near 0 indicate the lack of a linear correlation. These are illustrated in Figure 4.1.

If the relation between x and y is nearly linear, then the value of r measures the strength of the correlation. As $|r|$ decreases from 1 toward 0, the correlation weakens; this is illustrated in Figure 4.2. However, when $r = 0$ we can only conclude that there is no linear correlation; x and y might be strongly correlated by some nonlinear relation, as is

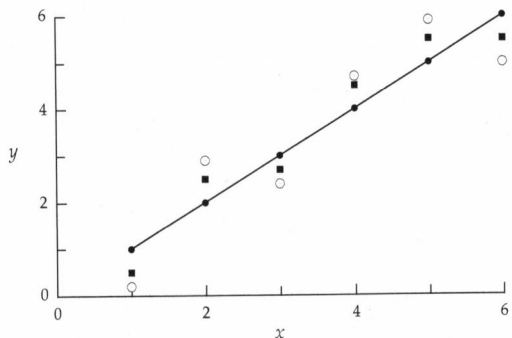

FIGURE 4.2   Strength of linear correlation as measured by the linear correlation coefficient r. Closed circles have $r = +1$; squares have $r = 0.96$; open circles have $r = 0.90$.

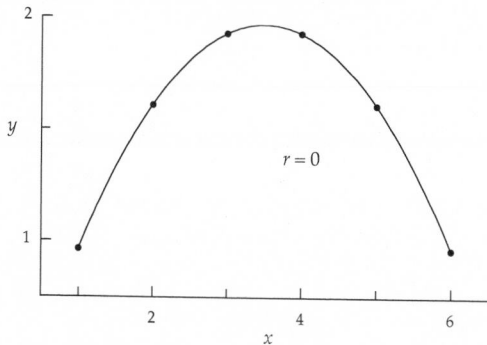

FIGURE 4.3 These six points are exactly correlated by a quadratic polynomial, even though the linear correlation coefficient $r = 0$.

shown in Figure 4.3. In other words, r can only measure the strength of linear correlations; if your data are nonlinear, then try linearizing the data (§ 3.1) before computing r.

The lesson here is that if, before performing the experiment, you already know that a significant linear correlation exists between x and y, then the linear correlation coefficient r is a good way to measure its strength. However, if you do not know whether x and y are correlated, then r is a poor way to test for a correlation [1].

4.2   Rank-Order Correlation Coefficients

Rather than use the linear correlation coefficient to test for a correlation, we advocate ranking the measured x and y values and then testing for a correlation between the two rankings. The test is applied by computing the rank-order correlation coefficient ρ. For a set of N pairs of measured (x, y) values, the procedure is as follows:

(a) Sort the x_i in either increasing or decreasing order and assign a rank R_i to each, $i = 1, 2, \ldots, N$.

(b) Likewise, sort the y_i in the same order and assign a rank Q_i to each of them.

(c) If any two or more x (or y) values are the same, give each the same rank computed as the mean of the ranks they would have if their values differed slightly [1, 2]. This mean will be either an integer or a half integer. For example, consider the five x values

TABLE 4.1 Example of five x-values, two of which are the same, and their ranks R_k.

k	x_k	R_k
1	7	4
2	4	(2+3)/2 = 2.5
3	3	1
4	4	(2+3)/2 = 2.5
5	8	5

in Table 4.1. Of these five, the value $x = 4$ appears twice. Those two values should have taken ranks 2 and 3, but since they are the same, they are both given the mean rank 2.5. When ordered by rank, the values appear as in Table 4.2.

(d) To test for a correlation between the ranks of x and y, compute the rank-order coefficient,

$$\rho = \frac{\sum_i^N (R_i - R_m)(Q_i - Q_m)}{\sqrt{\left(\sum_i^N (R_i - R_m)^2\right)}\sqrt{\left(\sum_i^N (Q_i - Q_m)^2\right)}} \qquad (4.2)$$

Here R_m is the mean of the x-ranks R_i and Q_m is the mean of the y-ranks Q_i. These two means are always the same

$$Q_m = R_m = \frac{N+1}{2} \qquad (4.3)$$

(e) When the ranks R and Q exhibit a strong positive linear correlation, then $\rho \approx 1$. When they exhibit a strong negative correlation, then $\rho \approx -1$. However, when there is no linear correlation between the ranks R and Q, then $\rho = 0$.

TABLE 4.2 Rank ordering of the five x-values from Table 4.1.

R_k	1	2.5	2.5	4	5
x_k	3	4	4	7	8

One advantage to using ρ instead of the linear coefficient r is that, for ρ to apply, the relation between x and y need only be monotonic; it need not be linear. That is, linearity is required between the ranks R and Q, not between x and y. This means that ρ is more general than r.

A second advantage is that ρ is a more robust measure of correlation than is r: it is less sensitive to uncertainties in values measured for x and y. This is because the ranks are less sensitive. Consequently, the rank-order coefficient ρ is often unaffected by small statistical errors or modest changes in x or y [1]. Even systematic errors in x and y may not affect the rankings R and Q.

A third advantage is that we know the distributions for the ranks R and Q [1]: each is uniform on $[1, N]$. Therefore we can assign a significance to our computed ρ. If the significance is 5%, then the correlation between x and y is said to be significant; if it is 1%, then the correlation is said to be highly significant. The latter means that if the experiment were repeated 100 times, then in only one of those experiments would chance be responsible for the value computed for ρ. The significance depends on the number of measurements N, as well as on the value of ρ. Table 4.3 gives the minimum values for ρ needed for 5% and 1% significance.

TABLE 4.3   Minimum values for ρ at 5% and 1% significance. Taken from [3].

N	5%	1%
5	1.0	...
6	0.89	1.0
7	0.79	0.93
8	0.74	0.88
9	0.68	0.83
10	0.65	0.79
12	0.59	0.78
14	0.54	0.72
16	0.51	0.66
18	0.48	0.62
20	0.45	0.59
25	0.40	0.53
30	0.36	0.48

The principal disadvantage to using ρ is that we lose information when we replace x and y with their ranks [1]. Specifically, from ρ we can only learn whether or not x and y are correlated monotonically. If they are correlated, the ranks cannot help us identify a function that could model the relation. Nevertheless, in many situations, simply knowing whether or not a correlation exists can be valuable.

4.3 Sample Calculation

We illustrate use of the rank-order coefficient by asking whether the grades in one course correlate with grades in a previous course. We have grades for twelve students from a fall course and, for the same students, from another course taken the following spring. These data are given in Table 4.4; they are plotted in Figure 4.4.

Using (4.1), we compute the linear correlation coefficient for these twelve pairs of points and find $r = 0.64$. From this value we are hard pressed to say whether or not the grades are correlated. Figure 4.4 suggests that a correlation exists, but it is probably nonlinear; if it is nonlinear, then the linear correlation coefficient cannot be used to measure the strength of the correlation.

So we turn to the rank-order coefficient. We rank the grades in the two courses, obtaining Table 4.5. Note in Table 4.4 that three students

TABLE 4.4 Final grades for twelve students in two courses.

Student	Fall Grade	Spring Grade
A	50.4	30
B	67.2	57
C	55.3	34
D	52.5	41
E	64.1	46
F	60.9	43
G	64.8	43
H	58.4	43
J	54.8	24
K	60.9	38
L	66.8	62
M	40.0	38

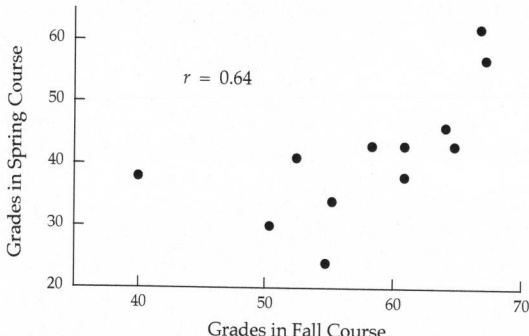

FIGURE 4.4 Grades from Table 4.4 for twelve students in a fall course and in a spring course. The linear correlation coefficient is 0.64.

have the same grade of 43 in the spring term; these should fill ranks 4, 5, and 6, whose mean is 5. Therefore, the rank $Q = 5$ appears three times in Table 4.5.

The ranks in Table 4.5 are plotted in Figure 4.5. Using these ranks in (4.2), we compute the rank-order coefficient and find $\rho = 0.81$. From Table 4.3 we see that for $N = 12$ points, we must have $\rho \geq 0.78$ for 1% significance. Therefore we conclude that, with a high degree of confidence, the grades for these students in these two courses are correlated.

TABLE 4.5 Rankings of student grades from Table 4.4.

Student	R(Fall)	Q(Spring)
B	1	2
L	2	1
G	3	5
E	4	3
F	5.5	5
K	5.5	8.5
H	7	5
C	8	10
J	9	12
D	10	7
A	11	11
M	12	8.5

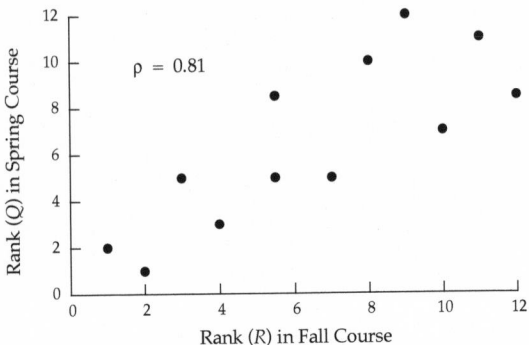

FIGURE 4.5 Ranks of student grades in Table 4.4. The rank-order correlation coefficient (4.2) is computed to be $\rho = 0.81$, which has a significance of better than 1%.

This suggests (but does not prove) that performance in the first course affects performance in the second. However, this result tells us nothing about any correlation of grades for other students. Nevertheless, the existence of a correlation for certain students might be important to those students and their instructors, even though we cannot say what form the correlation takes.

4.4   Cautions

We close this chapter by warning you that even if you can establish the existence of a correlation, you have not established a complete causal connection. The typical possibilities are these [4]:

(a) Your correlation between x and y might be the result of chance. Even with 1% significance, there is still one experiment out of 100 in which chance causes the apparent correlation. This eventuality is most common with small amounts of data, that is, when $N < 10$.

(b) Even when the correlation is real, there may not be a causal connection. For example, two clocks may strike at the same time (correlation), but this does not mean that the striking of one caused the striking of the other. To establish a causal connection you must develop and test a chain of events or circumstances that connect effect to cause. The incentive to seek such chains of logic may be bolstered by knowing that a correlation exists.

(c) Even when x and y are both correlated and causally connected, the rank-order correlation coefficient does not tell us whether x causes y or y causes x. We need additional information. Usually the information is temporal: part of the distinction between cause and effect is that the cause happens before the effect. This seems simple, but there are situations in which the temporal order of events is obscure. Examples include biological activities at the molecular and cellular levels.

Literature Cited

[1] W. H. Press, B. P. Flannery, S. A. Teukolsky, and W. T. Vetterling, *Numerical Recipes*, Cambridge University Press, Cambridge and New York, 1986.

[2] J. R. Taylor, *An Introduction to Error Analysis*, University Science Books, Mill Valley, CA, 1982.

[3] B. J. Underwood, C. P. Duncan, J. T. Spence, and J. W. Cotton, *Elementary Statistics*, Appleton-Crofts, New York, 1954.

[4] D. Huff, *How to Lie with Statistics*, Norton, New York, 1954.

[5] J. W. Tukey, *Exploratory Data Analysis*, Addison-Wesley, Reading, MA, 1977.

Exercises

4.1 Use the data in Table 4.5 to test (4.3); that is, determine whether the values of R and Q in Table 4.5 have the same values for their means.

4.2 The question has arisen as to whether y must be linearly related to x in order for the rank-order coefficient to be unity ($\rho = 1$). Test this using the quadratic $y = x^2$ with $x = 1, 2, 3, 4, 5, 6$. Explain your conclusion.

4.3 Determine the ranks (ascending order) for the following five x-values: 10, 12, 10, 12, 10.

4.4 Does the value of the rank-order coefficient ρ depend on whether the data are sorted in ascending or descending order? Justify your answer.

4.5 If the ranks R, arranged in ascending order, correlate with the ranks Q, arranged in descending order, near what value does the correlation coefficient ρ lie?

4.6 The following table gives the number of marriages and number of motor-vehicle deaths for six states in 1958:

State	# Marriages	# MV Deaths
NM	5,850	408
CN	16,879	251
KS	15,481	554
AL	24,444	852
OK	33,444	667
MA	45,959	590

Determine whether the number of marriages correlates with the number of deaths. (Data from Tukey [5].)

4.7 Now add the following four states to the table in the previous problem and test whether the numbers of deaths and marriages for these ten states are correlated. Compare your result with that found in the previous problem and discuss.

State	# Marriages	# MV Deaths
WY	2,945	137
GA	45,863	956
IA	25,101	598
CO	14,688	396

4.8 The rank-order correlation coefficient does not apply if the relation between the ranks of x and y is nonlinear. Give examples of relations between x and y that produce nonlinear relations between their ranks.

4.9 Here are vapor pressures for water at selected temperatures (see Figure 3.5):

$T(°C)$	30	100	150	200	250	300
$P^s(\text{bar})$	0.042	1.014	4.762	15.55	39.76	85.88

(a) From these data compute the linear correlation coefficient, r. Based on your value for r, what can you say about a possible correlation between these temperatures and the corresponding vapor pressures?

(b) Now make a table of the ranks from the above data. *Without doing any further calculations,* can you give the value for the rank-order correlation coefficient, ρ? From your value, what can you conclude about any correlation between these temperatures and vapor pressures? Be specific; for example, can you assign a significance to your value for ρ?

4.10 (a) Determine the linear correlation coefficient, the rank-order correlation coefficient, and its significance for the following data:

x	1	2	3	4	5	6
y	1.1	2.3	2.1	2.8	3.7	3.5

(b) After the data in (a) were collected, a calibration error was discovered; the correct data are these:

x	1	2	3	4	5	6
y	0.9	2.5	2.4	2.7	3.6	3.4

Determine the linear correlation coefficient r, the rank-order correlation coefficient ρ, and its significance for the these data. What do you conclude on comparing the values for r and ρ obtained in (a) and (b)?

4.11 (a) Is it possible for a set of x-y data to have the same value for the linear and rank-order correlation coefficients? That is, can a particular set of data have $r = \rho \neq 0$? If so, give an example for $N = 5$ x-y pairs.

(b) Is it possible for a set of x-y data to have $|\rho| > |r|$? If so, give an example for $N = 5$ x-y pairs; if not, explain why not.

(c) Is it possible for a set of x-y data to have $|\rho| < |r|$? If so, give an example for $N = 5$ x-y pairs; if not, explain why not.

4.12 Measurements of y at selected x-values gave the following data, along with the relative uncertainties in the measured ys:

x	y	u_y/y
1	12.7	1.6%
2	29.2	0.7%
3	34.8	0.6%
4	40.5	0.5%
5	49.3	0.4%

(a) A colleague claims that these results signify the presence of a systematic error that consistently increases as x decreases. To support this claim, he has computed the rank-order correlation coefficient between the y-values and their relative uncertainties given above. Repeat his calculation of ρ. Is systematic error the only plausible explanation for the resulting value found for ρ?

(b) From the values in the above table, compute the absolute uncertainties u_y, and then compute the rank-order correlation coefficient between the ys and these uncertainties. Compare this value of ρ with that found in (a) and explain any differences.

4.13 Write an essay of at least 500 words in which you explain, discuss, and critique the plot in Exhibit D.

4.14 For each pair of the following causes and their possible effects, label the connection between the two as either "unrelated," "correlated," or "causally connected." Give a brief rationale for each of your choices.

Proposed Cause	Possible Effect
1. Number of hours a student studies a course per week	1. Grade in course
2. Availability of internet in homes	2. Amount of housework performed by inhabitants
3. Number of female engineering professors	3. Number of female engineering students
4. Emissions from burning carbon fuels	4. Global warming
5. Fraction of personal income taxed by government	5. Amount of government revenues
6. Percentage of females in a class	6. Grades earned by males in the class
7. Amount of money taken in by a state lottery	7. Total monies available for that state's educational institutions.
8. Global warming	8. Number of Atlantic hurricanes
9. Number of laptop computers for middle-school students	9. Academic performance of students in high schools
10. Percentage of females in a class	10. Grades earned by females in the class
11. Strength of US economy	11. Amount of loose change on the asphalt of WalMart parking lots
12. Number of private jets in flight over US on a given day	12. Number of commercial jet flights delayed from their schedules that day

Exhibit E

Data from twelve laboratories for thermal conductivities of pure aluminum at cryogenic temperatures. All samples were judged to be at least 99.9% pure, so we expect all these points to fall on the same curve. However, at these low temperatures, impurities of only a few parts per thousand can change the thermal conductivity by as much as a factor of three.

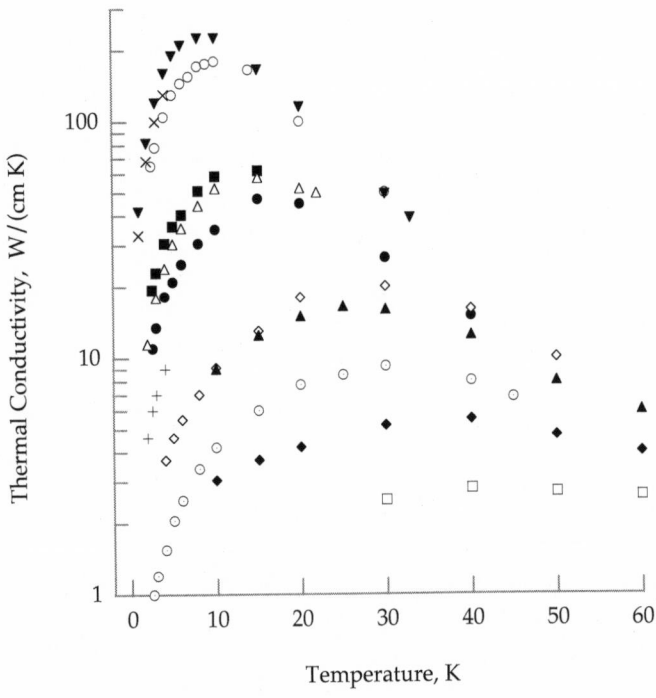

Data extracted from a more elaborate figure in Y. S. Touloukian, "The Impact of Physical Property Research on Technological Advancement," *Proceedings of the Seventh Symposium on Thermophysical Properties*, ASME, New York, 1977, p. 2.

5

Testing for Sensitivity

U P TO NOW WE HAVE DISCUSSED DATA obtained from a model experiment in which we measure values of a dependent variable y at precisely controlled values of an independent variable x. But in practice, the variable x can be controlled only to within some tolerance δx. For example, x might be temperature T and y might be the vapor pressure P of a fluid; then the experiment is to measure P at selected values of T. In such experiments, T is usually controlled by placing the fluid sample in a heat bath. The sample is in thermal equilibrium with the bath, and we control the temperature of the bath using heating elements, a sensing mechanism, and a feedback circuit. When the bath starts to cool, the heating elements are turned on to restore the desired T.

The problem is that, in such an apparatus, the temperature can be controlled to only within δT of a set-point T: the tolerance might be $\delta T = 1$ K, or 0.1 K, or even 0.01 K. But in any case, the issue is to determine how variations in T affect the dependent variable P: How sensitive is P to small changes in T?

The answers to these kinds of questions are important to us both in the *design* of the experiment and in the subsequent *analysis* of data. During design we need to know how tightly we must control the temperature to achieve a desired level of accuracy in the measured pressures. During analysis we assign uncertainties to the measurements, and for measured pressures, part of the uncertainty is caused by variations in temperature.

We start the presentation by defining common measures for sensitivity (§ 5.1) and showing how these definitions can be applied (§ 5.2). Then we extend the discussion to problems of stability in the control of steady-state flows (§ 5.3).

5.1   Sensitivity Measures

Let y represent a property of the experimental system. The value for y depends on the values of N other properties, x_i,

$$y = f(x_1, x_2, \ldots, x_N) \tag{5.1}$$

This is a measurement equation for the property y. Now define the *differential sensitivity* of y with respect to x_i to be the partial derivative

$$\left(\frac{\partial y}{\partial x_i}\right)_{x_j \neq x_i} = \lim_{\Delta x_i \to 0} \left(\frac{\Delta f}{\Delta x_i}\right)_{x_j \neq x_i} \tag{5.2}$$

The subscript $x_j \neq x_i$ means that the values of all xs are held fixed except that for x_i. So, the differential sensitivity tells how y responds to differential changes in one x when all others are constant. A large value for the derivative in (5.2) indicates high sensitivity: a small change in x_i causes a large response in y, as in region A of Figure 5.1.

In some cases it is helpful to have a *fractional* or relative sensitivity, which is defined by

$$\left(\frac{\partial \ln y}{\partial \ln x_i}\right)_{x_j \neq x_i} = \lim_{\Delta x_i \to 0} \left(\frac{\Delta \ln f}{\Delta \ln x_i}\right)_{x_j \neq x_i} \tag{5.3}$$

This allows us to determine the percentage response of y to a given percentage change in x_i. Note that the differential sensitivity, given by (5.2), has dimensions of y/x; in contrast, the fractional sensitivity, given by (5.3), is dimensionless.

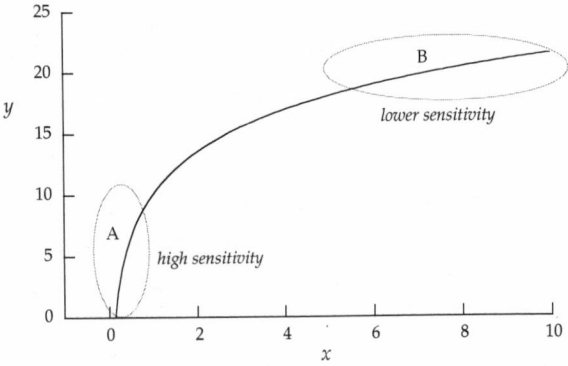

FIGURE 5.1 Sensitivity of y to a change in x usually differs over different ranges of x. For example, in region A of this plot, y is highly sensitive to x: small changes in x produce a large change in y. But in region B, y exhibits lower sensitivity: small changes in x produce small changes in y.

Let Δx_i represent a small finite change in x_i; we call this an incremental change. Then Δy represents the incremental response of y, given by

$$\Delta y = \left(\frac{\partial f}{\partial x_i}\right)_{x_j \neq x_i} \Delta x_i \qquad (5.4)$$

If all the xs change incrementally, then the total response of y is

$$\Delta y = \sum_i^N \left(\frac{\partial f}{\partial x_i}\right)_{x_j \neq x_i} \Delta x_i \qquad (5.5)$$

Let y_0 be the value for y when all the xs have values x_{i0}, then the value of y after an incremental change is, from (5.5),

$$y = y_0 + \sum_i^N \left(\frac{\partial f}{\partial x_i}\right)_{x_j \neq x_i} \Delta x_i \qquad (5.6)$$

This is the Taylor expansion for y, truncated at first order.

We caution that in applying these definitions, you must be aware of constraints imposed either by the experimental design or by nature. For example, if a fluid completely fills a rigid vessel, then the experimental design imposes a constant-volume constraint. So you can only evaluate constant-volume measures of sensitivity. Moreover, nature imposes constraints via conservation laws, such as conservation of mass and energy. For example, for a binary mixture, you cannot ask about sensitivity to changes in one mole fraction with the other held fixed.

5.2   Sample Calculations

We present two simple applications of the definitions from § 5.1. The first (§ 5.2.1) involves vapor pressure—a quantity that depends on only one variable. The second (§ 5.2.2) uses mole fractions of a binary mixture, and therefore involves a quantity that depends on two variables.

5.2.1   Vapor Pressures

To illustrate the definitions in the previous section, consider vapor pressures of water, which depend only on temperature. In Figure 3.13 we showed that a plot of $\ln(P^s)$ vs $1/T$ gives nearly a straight line. To improve on that representation, we add a quadratic term, so our correlation becomes

$$\ln P^s \ = \ A \ - \ \frac{B}{T} \ + \ \frac{C}{T^2} \tag{5.7}$$

For water, $A = 11.762$, $B = 3801.7$, and $C = -2.186(10^5)$; these apply for T in K and P^s in bar. At $100°C$ the correlation (5.7) gives $P^s = 1.004$ bar, which is below the true value (1.013 bar) by less than 1%.

Using (5.7) in (5.2), we find the vapor pressure of water to have a differential sensitivity that depends on temperature:

$$\frac{dP^s}{dT} \ = \ \frac{P^s}{T^2} \left(B \ - \ \frac{2C}{T} \right) \tag{5.8}$$

At 300 K, (5.8) gives

$$\frac{dP^s}{dT} \ = \ \frac{0.002 \text{ bar}}{K} \tag{5.9}$$

That is, the vapor pressure increases by 0.002 bar in response to a 1 K increase in temperature. Although this appears small, (5.7) gives the vapor pressure at 300 K as 0.035 bar; so 0.002 bar represents almost a 6% increase. In a design situation, this means that if we want to measure P^s to within ±1%, then we must control T to about 300.0 ± 0.2 K. In an analysis situation, (5.9) means that random temperature fluctuations of ±0.2 K contribute statistical errors of ±1% to the pressure measured at 300 K.

At 400 K, the differential sensitivity increases to

$$\frac{dP^s}{dT} \ = \ \frac{0.075 \text{ bar}}{K} \tag{5.10}$$

But this corresponds to a 3% increase in P^s for a 1 K increase in T. So to measure P^s to within ±1%, we would need to control the temperature to 400.0 ± 0.3 K.

From (5.9) and (5.10) we see that the differential sensitivity of the vapor pressure increases with temperature; in contrast, the fractional sensitivity decreases. From (5.8) the fractional sensitivity is given by

$$\frac{d \ln P^s}{d \ln T} \ = \ \frac{1}{T} \left(B \ - \ \frac{2C}{T} \right) \tag{5.11}$$

At 300 K we find the fractional sensitivity from (5.11) to be 17.5; that is, if the temperature increases by 1% (from 300 K to 303 K), then the vapor pressure increases by 17.5%. However, at 400 K the fractional sensitivity from (5.11) drops to 12.2. Now, if the temperature increases by 1% (from 400 K to 404 K), then the vapor pressure will increase by only 12.2%.

5.2.2 Mole Fractions of a Binary Mixture

As a second example, consider mole fractions of a binary mixture of components A and B. To determine the mole fractions, we intend to measure the number of moles n_A of component A and the number of moles n_B of component B. Then we will compute the mole fraction of A as

$$x_A = \frac{n_A}{n} = \frac{n_A}{n_A + n_B} \tag{5.12}$$

where n is the total number of moles.

We ask how the mole fraction responds to changes in n_A and n_B. Applying (5.2) to (5.12), the differential sensitivity of x_A to changes in n_A is

$$\left(\frac{\partial x_A}{\partial n_A}\right)_{n_B} = \frac{1 - x_A}{n} = \frac{x_B}{n} \tag{5.13}$$

Similarly,

$$\left(\frac{\partial x_A}{\partial n_B}\right)_{n_B} = \frac{-1 + x_B}{n} = -\frac{x_A}{n} \tag{5.14}$$

First note from (5.13) and (5.14) that the sensitivity decreases with increasing n; therefore, one way to improve accuracy is to increase the size of the sample mixture. Second note from (5.13) that x_A is most sensitive to changes in n_A when $x_B \approx 1$; that is, when the mixture is dilute in component A. Third, from (5.14) x_A is most sensitive to changes in n_B when $x_A \approx 1$; that is, when the mixture is dilute in component B.

Combining (5.13) and (5.14) into (5.5), we can write the total incremental change in x_A as

$$\Delta x_A = \frac{x_B}{n}\Delta n_A - \frac{x_A}{n}\Delta n_B \tag{5.15}$$

When changes in both n_A and n_B occur near the equimolar composition ($x_A = x_B$), the two terms in (5.15) tend to cancel. Consequently, mole fractions obtained from measurements of n_A and n_B will be most accurate for large quantities of equimolar mixtures and least accurate for small quantities of dilute solutions. This behavior is common: many quantities are most reliably measured in the middle of their ranges and less reliably measured near their extremes.

From (5.13) we find the fractional sensitivity of x_A to changes in n_A to be given by

$$\left(\frac{\partial \ln x_A}{\partial \ln n_A}\right)_{n_B} = x_B \qquad (5.16)$$

An analogous equation can be obtained from (5.14) for the fractional sensitivity of x_A to changes in n_B. Equation (5.16) shows that the fractional sensitivity is independent of the total amount of material n; however, like the differential sensitivity (5.13), the fractional sensitivity for x_A is largest for solutions dilute in component A.

5.3   Stability of Steady-State Flows

To illustrate sensitivity of systems in steady-state flows, we consider a simple mixing tank. We show how steady-state material balances can be used to analyze for sensitivity to disturbances (§ 5.3.1), then provide a numerical example (§ 5.3.2).

5.3.1   Material Balances for Steady-State Mixing

Consider a tank, as in Figure 5.2, which combines two feed streams and discharges a single product stream. Each stream is a binary mixture of components A and B. Feed stream 1 has molar flow rate n_1 and component A mole fraction x_1. Similarly, feed stream 2 flows at n_2 with mole fraction x_2. The discharge flows at rate n_o with mole fraction x_o.

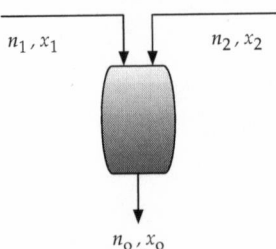

FIGURE 5.2 Schematic of a mixing tank that mixes two mixtures of components A and B; n_i are molar flow rates and x_i are mole fractions of component A.

At steady state, the total number of moles fed to the tank must be balanced by the total moles discharged,

$$n_1 + n_2 = n_o \tag{5.17}$$

and the total moles of A fed must be balanced by the total moles of A discharged,

$$x_1 n_1 + x_2 n_2 = x_o n_o \tag{5.18}$$

In the typical material balance problem, we know the feed compositions (x_1 and x_2), and we need to find the required feed rates (n_1 and n_2) that provide a desired product composition and flow rate (x_o and n_o). Then (5.17) and (5.18) can be solved for the required unknowns.

Note that (5.17) and (5.18) are linear in the unknowns, so they can be written in matrix form,

$$\mathbf{M n} = \mathbf{b} \tag{5.19}$$

Here $\mathbf{n}$ is the vector of unknowns,

$$\mathbf{n} = \begin{pmatrix} n_1 \\ n_2 \end{pmatrix} \tag{5.20}$$

$\mathbf{b}$ is the vector of right-hand sides,

$$\mathbf{b} = \begin{pmatrix} n_o \\ x_o n_o \end{pmatrix} \tag{5.21}$$

and $\mathbf{M}$ is the coefficient matrix,

$$\mathbf{M} = \begin{bmatrix} 1 & 1 \\ x_1 & x_2 \end{bmatrix} \tag{5.22}$$

Formally, the solution to (5.19) is obtained by inverting the coefficient matrix,

$$\mathbf{n} = \mathbf{M}^{-1} \mathbf{b} \tag{5.23}$$

But rather than solve these for the unknown flow rates $\mathbf{n}$, we are interested in a sensitivity analysis: in response to small changes in either

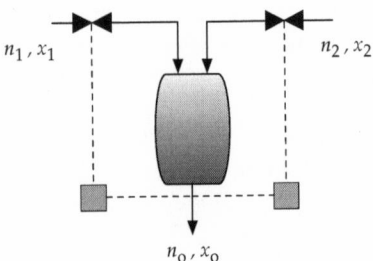

FIGURE 5.3   Same mixing system as in Figure 5.2, but with controllers added to sample the product stream and adjust valves on the feed streams to maintain the product composition and flow rate.

the feed compositions or the output flow, how must the flow rates be adjusted to maintain the desired output compositions? This is a problem of control; so, to the basic diagram in Figure 5.2, we add controllers that sample the product stream and manipulate a valve on each feed stream to maintain the desired product. This is illustrated schematically in Figure 5.3.

When changes occur in the product stream, the controllers manipulate the feed streams and the original problem (5.19) becomes

$$(\mathbf{M} + \delta\mathbf{M})\,(\mathbf{n} + \delta\mathbf{n}) \;=\; \mathbf{b} + \delta\mathbf{b} \tag{5.24}$$

The solution to this, analogous to (5.23) is

$$\mathbf{n} + \delta\mathbf{n} \;=\; (\mathbf{M} + \delta\mathbf{M})^{-1}\,(\mathbf{b} + \delta\mathbf{b}) \tag{5.25}$$

The feed rates have low sensitivity when small changes $\delta\mathbf{M}$ and $\delta\mathbf{b}$ (caused by small changes in the product stream) require only a small response $\delta\mathbf{n}$ (imposed by small adjustments in the feed streams). However, when small changes require a large response $\delta\mathbf{n}$, then the system exhibits high sensitivity. Equation (5.25) shows that the response $\delta\mathbf{n}$ is related to the inverse of the coefficient matrix; therefore, we expect high sensitivity to occur when the determinant of the coefficient matrix is close to zero,

$$|\mathbf{M}| \;=\; 0 \tag{5.26}$$

In such situations it is difficult (or impossible) to adjust the feed rates so as to maintain the required output; consequently, it becomes difficult (or impossible) to obtain accurate measurements of compositions and flow rates.

5.3.2 Numerical Example

To illustrate numerically, let's assign some values to the quantities in Figure 5.3. Let us require the output to be equimolar ($x_o = 0.5$) at $n_o = 20$ mol/s. The compositions of the feeds are initially $x_1 = 0.49$ and $x_2 = 0.51$. Then the material balances (5.19) require the steady-state feed rates to be $n_1 = n_2 = 10$ mol/s.

Now consider a disturbance that changes, by a small amount ε, the composition of feed stream 2:

$$x_2 \quad \rightarrow \quad x_2 + \varepsilon \qquad (5.27)$$

Our problem is to find how the feed rates must be adjusted to maintain the required output composition and flow rate.

Note the small value for the determinant of the initial coefficient matrix,

$$|\mathbf{M}| \quad = \quad \begin{vmatrix} 1 & 1 \\ 0.49 & 0.51 \end{vmatrix} \quad = \quad 0.02 \qquad (5.28)$$

So we anticipate a large response to the small change ε. Also note that in this problem, the coefficient matrix does not contain any flow rates: in this case, the sensitivity is determined solely by the mole fractions of the feed streams.

Replacing x_2 in (5.22) with the rhs of (5.27) and applying Cramer's rule, the new values of the feed rates are given by

$$n_1 \quad = \quad \frac{\begin{vmatrix} 20 & 1 \\ 10 & (0.51 + \varepsilon) \end{vmatrix}}{0.02 + \varepsilon} \quad = \quad 10 + \frac{10\varepsilon}{0.02 + \varepsilon} \qquad (5.29)$$

and

$$n_2 \quad = \quad \frac{\begin{vmatrix} 1 & 20 \\ 0.49 & 10 \end{vmatrix}}{0.02 + \varepsilon} \quad = \quad \frac{0.2}{0.02 + \varepsilon} \qquad (5.30)$$

When there is no disturbance, so $\varepsilon = 0$, (5.29) and (5.30) reduce to $n_1 = n_2 = 10$ mol/s, as they should.

Consider a 1% increase in x_2; that is, $\varepsilon = 0.005$, so $x_2 = 0.505$. Then (5.27) and (5.28) give

$$n_1 = 12 \text{ mol/s} \quad \text{and} \quad n_2 = 8 \text{ mol/s} \quad (5.31)$$

In response to a 1% increase in x_2, the feed rate of stream 1 must increase by 20% while that for stream 2 must decrease by 20%, if the discharge rate and composition are to remain the same. These large responses to a small disturbance would create difficulties in controlling the system and in measuring data from it. In some flow situations the determinant of the coefficient matrix may become small enough that the system is unstable: in response to a small disturbance, the system response becomes unbounded and therefore uncontrollable.

5.4   Comment

To apply the definitions of sensitivity given in § 5.1 we must have a theoretical context that relates the dependent variables to the independent ones. To the extent that the theoretical context is approximate, the corresponding estimates of sensitivity will be approximate. Alternatively, if the theoretical context fails to provide a measurement equation, we can estimate the sensitivity experimentally by imposing a small change Δx and measuring the response Δy. In either case, our estimates of sensitivity should be sufficient to guide us toward sound experimental designs and meaningful analyses. Further, they should help us avoid the frustrations of dealing with nearly unstable systems.

Exercises

5.1   The compressibility factor is defined by $Z = PV/nRT$, where the gas constant R = 0.08314 (liter bar)/(mol K).

   (a) For ten moles of gas in a one-liter vessel at 300 Kelvin and 20 bar, what is the sensitivity of Z to changes in T? What is the fractional sensitivity?

   (b) If P, V, n, and T all simultaneously increased by 1%, what would be the incremental change in Z?

   (c) Consider fixed V and n. Show whether or not P and T can change such that there is no incremental change in Z.

5.2 For binary ideal-gas mixtures of components 1 and 2, the heat capacity is the mole-fraction average of the heat capacities of the pure components:

$$C_{mix} = x_1 C_{pure1} + x_2 C_{pure2}$$

If $C_{pure1} = 5R/2$ and $C_{pure2} = 9R/2$, then determine the sensitivity of C_{mix} to changes in the mole fraction x_1. Here R is the gas constant, 8.314 J/(mol K) .

5.3 The mass flow rate g of fluid flowing through a pipe is to be measured by a bucket-and-scales method. A mass of fluid m is collected over an elapsed time t; then $g = m/t$.

(a) Show that the sensitivity of g to small variations in m depends only on the elasped time t.

(b) Show that the magnitude of the sensitivity of g to variations in t depends on both t and the flow rate g itself.

(c) From your results for (a) and (b) do you expect the most accurate values for g to be obtained from (i) high flow rates measured over long times, (ii) high flow rates over short times, (iii) low flow rates over long times, or (iv) low flow rates over short times?

(d) In a certain situation, we have a scale that can be read with an uncertainty of 0.1 lb_m and a stop watch that can be read with an uncertainty of 0.2 sec. For flows near 100 lb_m/min, what is the minimum duration over which water should be collected so that the scale and watch readings together contribute no more than 0.5% (0.5 lb_m/min) to the total uncertainty in the mass flow rate?

5.4 We need to determine the amount of heat Q required to raise the temperature of a mass of water m from 10°C to 90°C. The heat can be estimated by $Q = mC_p\Delta T$, assuming the heat capacity C_p remains constant when the temperature changes. This assumption introduces a small error: at 10°C, $C_p = 75.5$ J/(mol K), while at 100°C, $C_p = 76.0$ J/(mol K). If we can measure T within 0.2°C and m within 0.5 lb_m, what is the minimum mass of water that we should use so the total uncertainty (from m, T, and C_p) in Q is less than 1%?

5.5 The volume of a right cylindrical barrel is given by $V = \pi r^2 h$, where r is the radius of one circular face and h is the height of the barrel.

(a) The volume of a certain barrel is to be determined by measuring r and h. Will the value computed for V be more sensitive to the value measured for r or to that for h? For example, will a 1% error in the value measured for r, at fixed h, give a larger or smaller uncertainty in V than a 1% error in the value measured for h at fixed r.

(b) Consider two barrels, A and B. Both are known to have the same volume, but the radius of barrel A is much larger than that of barrel B. Would the volumes of the barrels be more reliably determined by measuring the radius and height of barrel A or of barrel B?

5.6 Two liquid-in-glass thermometers are exactly the same height and are filled with the same liquid. Each is calibrated to read from the freezing point to the boiling point of water; however, one is marked in increments of °F, while the other is marked in increments of °C. Can one of these thermometers give more precise values for temperature? Explain.

5.7 A mixing tank is being used to combine three ternary mixtures of components A, B, and C in steady-state flow. Let the feed streams be labeled 1, 2, and 3; let x_i be the mole fraction of A in stream i; let y_i be the mole fraction of B in stream i. Stream 1 has $x_1 = 0.20$, $y_1 = 0.10$; stream 2 has $x_2 = 0.22$, $y_2 = 0.12$; stream 3 has $x_3 = 0.70$, $y_3 = 0.30$. The tank output is to have $x_o = 0.55$, $y_o = 0.24$ at $n_o = 100$ moles/min.

(a) Write three independent material balances for the tank.

(b) Evaluate the determinant of the coefficient matrix from the equations in (a). Do you anticipate that the system will be sensitive to small disturbances in the feeds?

(c) Solve your equations for the feed rates n_1, n_2, and n_3.

(d) Now consider stream 3 becoming contaminated with a small amount of C. Specifically, the mole fractions in stream 3 change to $x_3 = 0.6975$, $y_3 = 0.2975$. Solve your material balances for this situation. Are the changes in n_1, n_2, and n_3 consistent with your answer to (b)?

5.8   The gas mileage m for an automobile can be expressed as $m = d/V$, where V is the volume of gasoline consumed when the vehicle has traveled distance d. For distances of around 400 miles (644 km) and volumes of about 15 gallons (56.8 liters), is the uncertainty in the computed mileage more sensitive to uncertainties in the odometer (which provides d) or to uncertainties in the meter on the gas pump (which provides V)?

5.9   The density of a certain liquid is known to be roughly 1.2 g/cm^3. The value is to be determined by separately measuring the mass m and volume V of a sample, then forming $\rho = m/V$.

(a) Show whether or not the reliability of the value obtained for ρ is affected by the amount of liquid used to obtain values for m and V.

(b) Show whether the fractional increment $\Delta\rho/\rho$ is more sensitive to uncertainties in m or in V. That is, should one of m or V be measured more reliably than the other?

5.10 Consider the steady-state mixing situation in Figure 5.3 in which stream 1 has $x_1 = 0.7$ and stream 2 has $x_2 = 0.2$. The output stream is to have $x_o = 0.25$ at a molar flow rate of $n_o = 50$ moles/min.

(a) Determine the required values for the feed flows n_1 and n_2.

(b) If x_2 increases by 2% to $x_2 = 0.204$, by what percentages must n_1 and n_2 change to preserve the desired output conditions?

(c) If x_2 increases by 30% to $x_2 = 0.26$, what can you say about the output stream?

5.11 Write an essay of at least 500 words in which you explain, discuss, and critique the plot in Exhibit E.

Exhibit F

When a biological sample is frozen and then thawed, its survival is affected by the freezing rate: the cooling rate must be neither too slow, nor too fast, but just right. Otherwise, so many cells are injured that the sample does not survive. As an example, the following plot shows measured percentages of surviving cells for samples of a certain bacterium frozen at different rates. Line is an unweighted least-squares fit. The location and height of the maximum varies with cell type, but all types show a maximum at some intermediate freezing rate. At slow cooling rates, most cells are injured by formation of ice within cells. At fast rates, icing of fluid between cells becomes lethal.

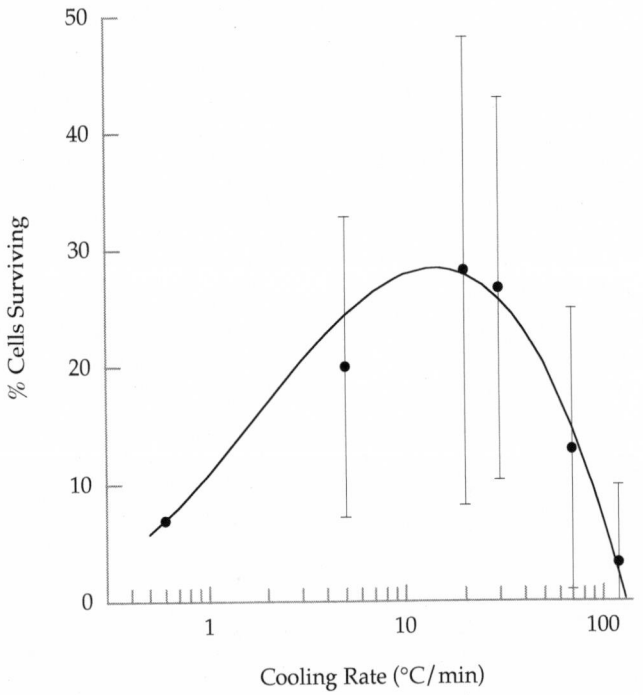

Plot redrawn from E. G. Cravalho, *Adv. Cryogenic Engr.*, 21, 399 (1975). Original reference does not show an error bar at a cooling rate of 0.6°C/min.

6

Attaching Meaning to Data

O NCE THE LABORATORY WORK IS COMPLETE and the data have been collected, then we arrive at the heart of the experiment: we must extract as much information as possible from the data and decide what that information means. These activities are not unlike those in detective fiction. The data provide evidence, but we (as the detectives) must interpret that evidence in such a way that we can construct a logical and self-consistent explanation of the events that gave rise to the evidence.

We divide the analysis into two parts: basic analysis (§ 6.1) and advanced analysis (§ 6.2–§ 6.6). The basic analysis is what Dr. Watson typically does; it must be done, but often it is all that gets done. When this happens, the experimentalist is failing to take full advantage of all the effort made in gathering the data. Our aim is to do more.

6.1 Basic Analysis

UNCERTAINTIES. The basic analysis begins with assessments of Type A and B uncertainties and the combining of those into a total uncertainty for each point (Ch. 2). By this process we double the amount of information available: at each measured point we have a value and an uncertainty. So when each point is plotted, we plot a value and an error bar. Thereafter, we always see the data together with their uncertainties.

INTERNAL CONSISTENCIES. Next we check the data for internal consistencies. These include verifications of applicable conservation laws, such as material balances and energy balances. These may apply to individual pieces of equipment or to the entire apparatus or both. If steady-state flows of mass and energy are assumed, these should be confirmed, to the extent possible.

Examples of other consistency checks include confirmation of directions of heat, mass, and momentum transfers in accordance with prevailing gradients. We should look for consistent changes in system parameters, such as temperatures, pressures, and compositions. For example, a heat exchanger might be used to cool a gas stream; then we should verify that the outlet temperature is, in fact, less than the inlet temperature.

THEORETICAL CONTEXT. At this point we try to connect the data to a theoretical context. We seek relations between dependent and independent variables using the methods of Chapter 3: linearization, least squares, and analysis of residuals. Then we place those relations within a theoretical context. We interpret the observed behavior within its theoretical context, show the extent to which the data conform to the theory, and explain the extent to which the data and theory fail to agree.

OTHER EXPERIMENTS. Lastly we mention the hazards involved in comparing data from different experiments, especially experiments performed in different labs. This activity has become a common part of basic analysis, but the comparisons reported are often so perfunctory that the subsequent conclusions are of little value. People usually think they are comparing apples with apples, when in fact they may be comparing apples with grapefruit (e.g., see Exhibits B and E). In some comparisons, we should be just as concerned when we obtain good agreement as when we don't.

To make such comparisons meaningful, you must be knowledgeable about both experiments: What is the same and what differs in the two procedures, in their limitations, in their theoretical foundations? Should you expect the data from the two to be comparable? If so, what uncertainties have been assigned to each set of data? Are the assigned uncertainties realistic? Remember you must compare not just two sets of data, but the data and their uncertainties.

If the comparison signals disagreement, can the disagreement be explained by differences in apparatus, procedure, analysis, theoretical context? If the comparison suggests agreement, is it real or is it the result of a fortuitous cancellation of errors? The number of variables and the possible explanations are many, making the investment in time nontrivial.

The basic analysis is what usually gets reported, so we will not belabor those activities further here. Instead, we want to emphasize how we can move beyond the basics. A systematic strategy for doing more would include these activities: extending important values to a range plus tolerance (§ 6.2), explaining extrema (§ 6.3), logically extending observations and conclusions (§ 6.4), specialization (§ 6.5), and generalization (§ 6.6). We consider each in turn.

6.2 Range Plus Tolerance

In most experiments, the central quantity of interest y is related to measurable quantities, x_i, through a measurement equation,

$$y = f(x_1, x_2, \dots) \tag{6.1}$$

In Chapter 5 we defined measures for the sensitivity of y to changes in any one or all of the xs. To apply those definitions we need some theoretical context to provide the functional relation between y and x. To the extent that the theoretical context is approximate, then the computed sensitivities are also approximate. So, experimental tests of the theoretical context should also include tests to confirm the anticipated sensitivity. This is particularly important when y is much more sensitive to one (or a few) of the xs than to the rest.

Sensitivity also enters the analysis when preparing conclusions. For example, consider an experiment intended to find the optimum efficiency of a piece of equipment. Analysis of the data might lead us to conclude that the efficiency is most sensitive to operating pressure. Then a study of the effects of operating pressure yields Figure 6.1, from which we may be tempted to report a statement like the following:

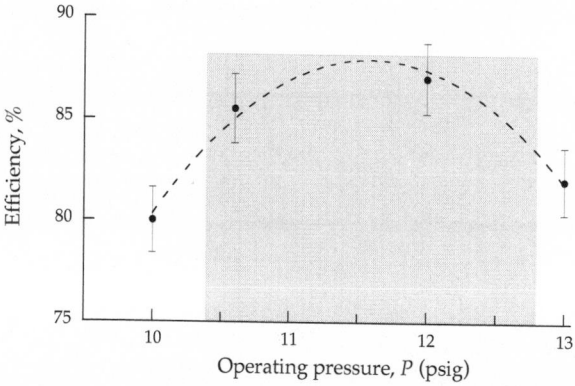

FIGURE 6.1 Effect of operating pressure on measured efficiencies of a device. Line is a quadratic fit to the four points. Maximum efficiency is 88% and occurs at 11.6 psi. Shaded region contains the range of pressures (± 1.2 psig) to operate within 5% of the optimum efficiency. Error bars are expanded uncertainties at the 68% confidence level.

When the operating pressure is 11.6 psig, then the device
operates at its optimal efficiency of 88%.

This is definitive, but it is less informative than it could be. In practice, we cannot operate continuously at precisely 11.6 psig; and so, what
we really need to know is how much uncertainty can be tolerated. In
fact, Figure 6.1 allows us to identify a range of operating pressures that
would be close to the maximum efficiency. We call this *range plus tolerance*; for example,

When the operating pressure is 11.6 ± 1.2 psig, then the
device runs within 5% of its optimal efficiency of 88%.

Now compare the behavior in Figure 6.1 with that in Figure 6.2;
both figures show the same optimum at the same operating pressure.
Nevertheless, the sensitivity to pressure in Figure 6.2 is higher than
that in Figure 6.1. Consequently, for the behavior in Figure 6.2, our
conclusion would be this:

When the operating pressure is 11.6 ± 0.6 psig, then the
device runs within 5% of its optimal efficiency of 88%.

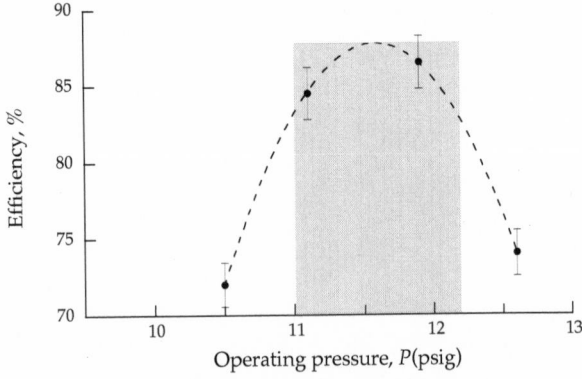

FIGURE 6.2   Effect of operating pressure on measured efficiencies of a device.
Line is a quadratic fit to the four points. Optimum efficiency is 88% and occurs at 11.6 psig. Shaded region gives the range of pressures (±0.6 psig) to operate within 5% of optimum efficiency; cf. Figure 6.1. Error bars are expanded
uncertainties at the 68% confidence level.

The range of operating pressures for Figure 6.2 is half the range for Figure 6.1. Reporting an operating *range* (which depends on sensitivity) plus a tolerance is often more valuable than reporting just an operating *point*.

Further note that we did not actually measure the optimum efficiency; we estimated it from an interpolation. And even if we had measured the optimum, it would have had some experimental uncertainty associated with it. Therefore, we could claim that reporting a range plus a tolerance better reflects the reality of the experimental situation than reporting just a point.

6.3   Explaining Extrema

To keep this book small, we have restricted our attention to linear or nearly linear relations between the measured variable y and the controlled variable x. Such situations are common, but also common are situations in which y exhibits an extremum—either a maximum or a minimum. When this occurs, we should seek explanations. Mathematically, there are many ways to explain an extremum, but simple ways include those in which y can be decomposed into two contributions that either reinforce one another or compete with one another. In such cases, extrema occur because the reinforcement or competition applies over different scales in x.

Consider a situation in which y decomposes into y_1 and y_2, with y_1 and y_2 reinforcing one another. In these cases an extremum may occur, and that extremum may be either a maximum or a minimum. For example, Figure 6.3 shows a minimum obtained from the sum of the

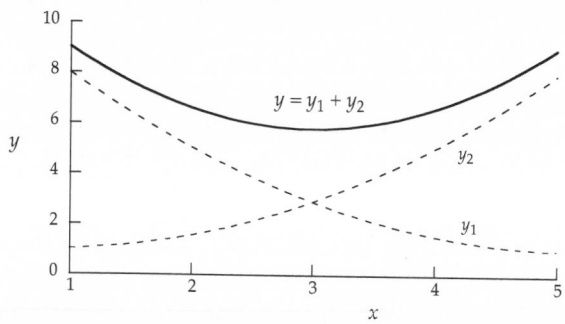

FIGURE 6.3   The sum of two monotonic, concave-up curves (y_1 and y_2) can produce a minimum.

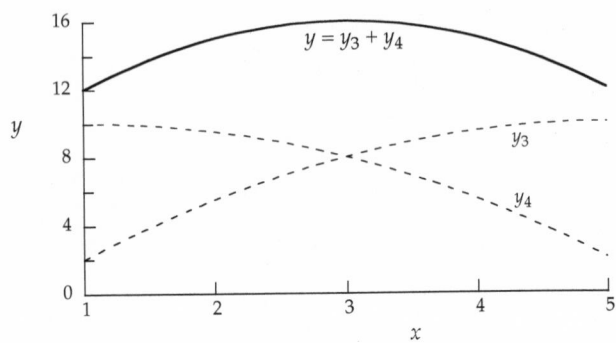

FIGURE 6.4   The sum of two monotonic, concave-down curves (y_3 and y_4) can produce a maximum.

two contributions, y_1 and y_2. Both y_1 and y_2 are monotonic, concave-up functions; however, y_1 decreases with x, while y_2 increases. The minimum occurs because y_1 dominates at small x while y_2 dominates at larger x. In contrast, Figure 6.4 shows a maximum from the sum of two other contributions, y_3 and y_4. Again, both y_3 and y_4 are monotonic functions, one increasing, the other decreasing with x, but both functions are concave down.

   Now consider situations in which the components of y compete with one another; here too extrema may be formed, and both maxima and minima are possible. For example, Figure 6.5 shows a minimum obtained from the difference between y_5 and y_4. Both functions

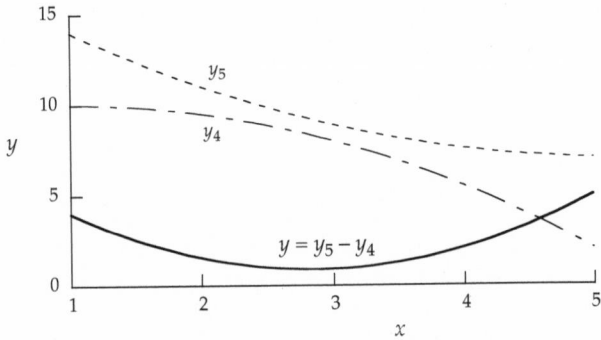

FIGURE 6.5   A difference between two monotonic curves, one concave up (y_5), the other concave down (y_4), can produce a minimum when the concave-up curve (y_5) lies above the concave-down curve (y_4).

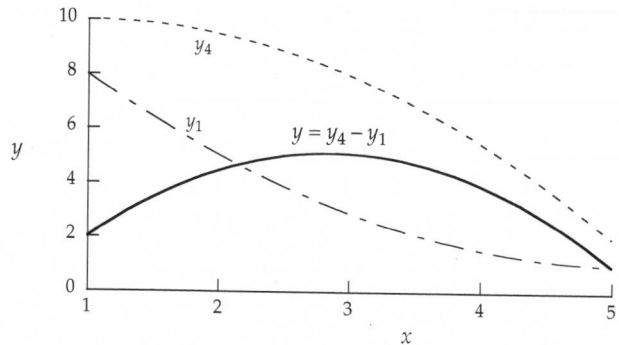

FIGURE 6.6 A difference between two monotonic curves, one concave up (y_1), the other concave down (y_4), can produce a maximum when the concave-up curve (y_1) lies below the concave-down curve (y_4).

are monotonic, but one is concave-up, while the other concave down; moreover, the concave-up function (y_5) lies above the concave-down function (y_4). The minimum occurs at an x value for which the competition between y_4 and y_5 is nearly balanced. Alternatively, a maximum can be formed from the difference of concave-up and concave-down functions, as in Figure 6.6. But now the concave-up function (y_1) lies below the concave-down function (y_4).

The lesson is that when data exhibit extrema, consider whether the behavior can be explained by decomposing y into two (or more) components whose contributions dominate y over different scales in x. Often those components will be monotonic, and then you might be able to linearize them using the procedure in § 3.1.

6.4 Logical Extensions

In summaries and conclusions drawn from experiments we continually make conditional statements of the form

$$\text{If } \mathcal{A}, \text{ then } \mathcal{B}. \tag{6.2}$$

For example, if the operating pressure is 11.6 psig, then the device runs at 88% efficiency. Once your analysis has established such relations, then you might be able to extract additional information by considering logical permutations. Simple permutations include the *converse*, the *inverse*, and the *contrapositive*; these are the three we discuss here.

Of these, the converse is the most abused and misused, the inverse is the one usually misunderstood, and the contrapositive is the one most neglected.

6.4.1   Converse

For the original conditional (If A, then B.), the converse takes this form:

$$\text{If } B, \text{ then } A. \tag{6.3}$$

The important point is that the truth of the converse (6.3) is not determined by the truth of the original condition (6.2): the converse may be true or false, regardless of the truth of the original statement (6.2).

For example, in calibrating a flow meter, we may establish experimentally that when the flow rate is 20 kg/s, then the meter reads 8.6 (arbitrary units). But when we use the calibration, we implement the converse: When the meter reads 8.6, then the flow rate is 20 kg/s. Calibrations are usually based on the truth of the converse.

As a counter example, consider this finding:

> If this liquid is ethanol (C_2H_5OH), then the molecular weight is 46.

This is a true statement. However, its converse,

> If the molecular weight of this liquid is 46, then the liquid is ethanol.

is false. Other substances, such as dimethyl ether (CH_3OCH_3), also have a molecular weight of 46.

A conditional and its converse are both true when there is a one-to-one correspondence between A and B. For example, when creating calibration curves, we try to establish a one-to-one correspondence between a measured value and a meaningful interpretation of that value. However, a statement and its converse will not have the same truth value if the correspondence between A and B is one-to-many.

Establishing the truth value of a converse may be important, regardless of whether it is true or false. If the converse is true, it may be useful in situations in which it is difficult to directly apply the original statement. When the converse is false, then it may be informative to discuss why it is false: what is the nature of the one-to-many relation?

Unfortunately, the original scope of an experiment may not be sufficient to determine the truth value of a converse: generally, the experimental design must include explicit provisions for exploring a converse.

6.4.2 Inverse

The inverse of the original conditional (If $\mathcal{A}$, then $\mathcal{B}$.) takes the form

$$\text{If not } \mathcal{A}, \text{ then not } \mathcal{B}. \qquad (6.4)$$

Like the converse, this statement may be true or false, regardless of the truth value of the original condition. From our calibration example, let's assume this statement is true by experiment:

> If the meter reads 8.6, then the flow rate is 20 kg/s.

Then the inverse would be

> If the meter does not read 8.6, then the flow rate is not 20 kg/s.

This inverse is also true because the calibration procedure establishes a one-to-one relation between the flow rate and the meter reading. However, although the following statement is true:

> If this is ethanol, then it has a molecular weight of 46.

it has a false inverse,

> If this is not ethanol, then it does not have a molecular weight of 46.

because the substance could have a molecular weight of 46 even though it is not ethanol.

The inverse provides another way to test for a one-to-many relation between $\mathcal{A}$ and $\mathcal{B}$. In some situations it can be more informative than the converse, such as cases in which $\mathcal{A}$ cannot be caused by $\mathcal{B}$. Most importantly, note that the inverse and converse have the *same* truth value: they are both true or both false. Therefore, if you can establish the truth value of an inverse, then you have also established the truth value of the converse.

6.4.3 Contrapositive

For the original conditional (If $\mathcal{A}$, then $\mathcal{B}$.), the contrapositive is

$$\text{If not } \mathcal{B}, \text{ then not } \mathcal{A}. \tag{6.5}$$

The contrapositive always has the *same* truth value as the original condition (6.2): they are both true or both false. Consequently, if you have, by experiment, established the truth of a condition, then you have also established its contrapositive. For example, this is a true statement:

If this liquid is ethanol, then the molecular weight is 46.

And its contrapositive is also true:

If the molecular weight is not 46, then this is not ethanol.

The contrapositive is always a useful test to apply to any experimental result; however, it may or may not provide any useful additional information. Note that the inverse is the contrapositive of the converse; see Table 6.1. We emphasize that while an experiment may establish the truth of a conditional (and its contrapositive), the data may or may not also establish the converse and inverse. They may be true or false; to decide, you may need additional data.

6.4.4 Empirical Inference

In a few domains of knowledge, things can be proved definitively; examples include geometry, number theory, and formal logic. However, most of our "knowledge" does not come from proofs; instead, it is inferred from observations. Science is the formalization of those activities

TABLE 6.1 Forms for three simple permutations of a conditional.

Name	Form	Truth Value
Original	If $\mathcal{A}$, then $\mathcal{B}$.	True by experiment
Converse	If $\mathcal{B}$, then $\mathcal{A}$.	True or false
Inverse	If not $\mathcal{A}$, then not $\mathcal{B}$.	Same as converse
Contrapositive	If not $\mathcal{B}$, then not $\mathcal{A}$.	Same as original

in which we infer relations from systematic experiments. But inferences are not proofs, and no matter how many experiments we perform nor how carefully we measure, we can never prove anything solely on the basis of experimental data. More data and more carefully measured data may strengthen some inferences, but more data may also falsify others [1]. Here is an example.

Figure 6.7 shows the time required to charge the battery on a laptop computer. Charging began when the battery was down to 14% of total charge, and the charge continued for 75 minutes. After 75 minutes, the battery had reached 75% of total charge. The points in Figure 6.7 suggest that the fraction of total charge is linear in the charging time; indeed, for those points, we find $r = 0.9996$. The solid line in the figure is a least-squares fit to the points. From these data, we might extrapolate the fitted line to estimate the time required to fully charge the battery. The extrapolation (broken line in the figure) gives a total charging time of 106 minutes.

To test our extrapolation, we continue to charge the battery beyond 75 minutes. These additional data are shown in Figure 6.8 where we see that, in fact, after 75 minutes the fraction of total charge is not linear in

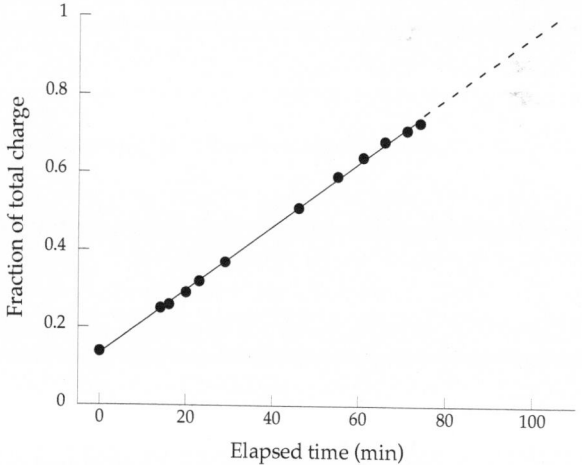

FIGURE 6.7 Time required to charge the battery on a laptop computer. Charging started when the battery was at 14% of total charge and continued for 75 minutes. Points are measured data and have $r = 0.9996$. Solid line is a least-square fit. Broken line is an extrapolation of the solid line. The extrapolation gives 106 minutes as the time required for the battery to reach total charge.

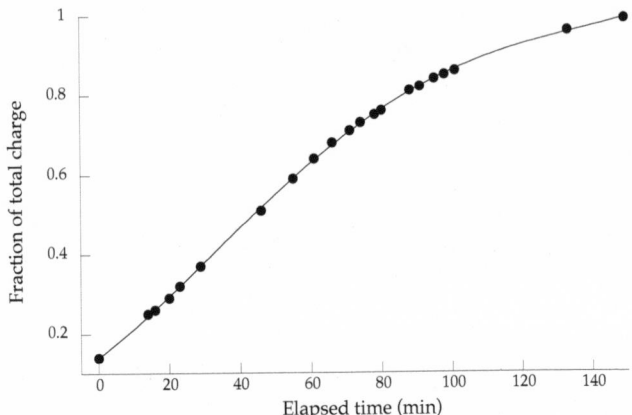

FIGURE 6.8 Time required to completely charge the battery on a laptop computer. For times < 75 minutes, points here are the same as those in Figure 6.7.
Solid line is a fourth-order polynomial fit to all the data.

the charging time. Instead, a fourth-order polynomial is needed to fit all
the data, as shown in Figure 6.8. And instead of 106 minutes, the battery
actually required 149 minutes to reach total charge. The extrapolation
of the straight line incurred a 29% error in the estimate for the total
charging time.

On the basis of the data in Figure 6.7, we *inferred* (we did not prove)
a linear relation between two variables. However, when more data were
collected (Figure 6.8), we found that our inference was mistaken: the relation between the two variables is more complicated. The lack of nonlinearities in the data in Figure 6.7 does not mean that nonlinearities
will not become apparent when the situation is studied more carefully.
Absence of evidence is not evidence for absence.

6.5   Specialization

By specialization we mean the use of data to study special cases. Usually, special cases are bounds or limiting behavior obtained when some
operating variable takes an extreme value—often either zero or infinity. Many times you cannot physically reach the extreme in the experiment, so you approach it by extrapolating the available data. In some
cases, the limiting special case has a known physical significance and
therefore it provides a consistency check on the data. In other cases,

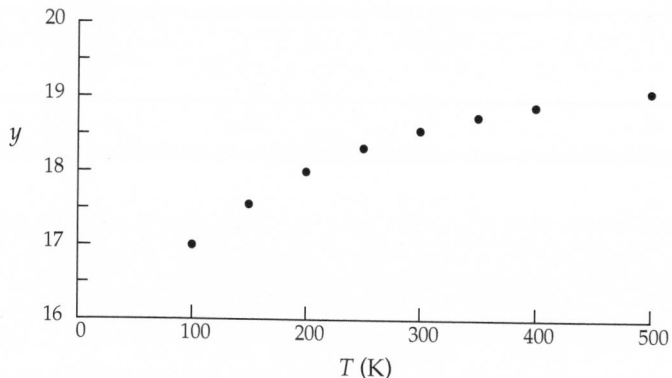

FIGURE 6.9 A quantity $y(T)$ whose limiting value we seek at high T.

the limit is a bound on behavior and it is instructive to know how that limit is approached and whether the experimental operating conditions are close to that bound.

When the special case is reached by driving an operating variable x to infinity, we rarely have enough data to extrapolate a y vs x plot. Common examples include infinite-time limits, infinite-temperature limits, and infinite-size limits. As a particular example, let us consider Figure 6.9, which shows a temperature-dependent variable $y(T)$. Assume a theoretical context suggests that the quantity y is bounded at high temperatures; we seek that bound by an extrapolation. But it is unlikely that we can reliably extrapolate the plot in Figure 6.9. In such cases, the infinite limit can usually be obtained by plotting against the reciprocal of the independent variable,

$$\lim_{x \to \infty} y(x) = \lim_{1/x \to 0} y(x) \tag{6.6}$$

Even if y vs $1/x$ is nonlinear, the behavior is often linear near $1/x = 0$. Then by fitting a straight line to a few points near $1/x = 0$, we can extrapolate with confidence to the limiting value. This is illustrated in Figure 6.10.

Some extrapolations lead to an idealized model of behavior; in such cases, extrapolation of the data serves both as a consistency check and as a way to measure how far removed the data are from the idealization. For example, consider the compressibility factors for steam plotted in Figure 6.11. At this temperature ($300°C$) and high densities, the data give $Z \ll 1$, indicating that steam is highly nonideal. Nevertheless, we

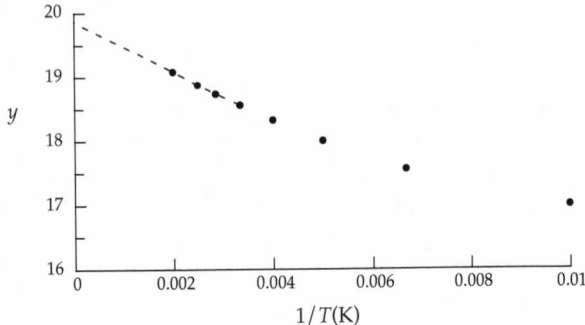

FIGURE 6.10 Same data as in Figure 6.5, but plotted vs $1/T$, so the infinite-temperature limit occurs at $1/T = 0$. Dashed line is a least-squares fit to the last four points; i.e., those at $T = 300$, 350, 400, and 500 K.

expect steam will approach ideal-gas behavior as the molar density decreases to zero. This means that at fixed temperature we expect

$$\lim_{\rho \to 0} (\text{real gas}) = \text{ideal gas} \tag{6.7}$$

Figure 6.11 shows this, increasing our confidence in the plotted points.

We routinely use many idealized models, such as ideal solutions, incompressible fluids, perfect crystals, laminar flows, steady states, complete mixing, first-order kinetics, perfect insulation, reversible changes of state, fully developed flows. All experiments invoke some idealized

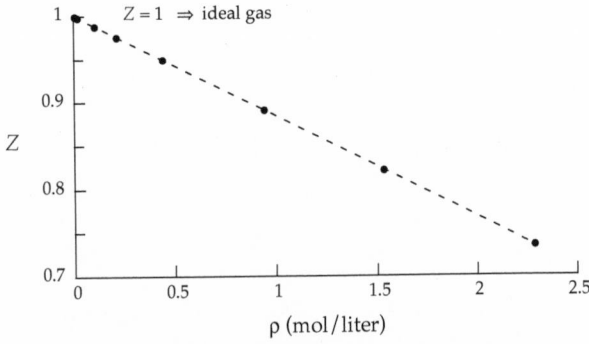

FIGURE 6.11 Compressibility factors Z for steam at $300°C$. Dashed line is a least-squares fit to the points. Extrapolation to $\rho = 0$ produces the ideal-gas value, $Z = 1$. Original data from NIST Webbook [2].

models, and we should develop the habit of using our data to test the extent to which the assumed models actually apply.

6.6 Generalization

When we generalize, we try to extend our results beyond the confines of our particular experiment. Polya [3] identifies two types of generalization: dilution and condensation. *Dilution* is easy, involves little thought, and is rarely valuable. In extreme cases, the generalizations are so dilute that they could have been stated *before* the experiment was done. For example, (a) to reduce heat losses, use smaller driving forces; (b) to increase the rate of heat transfer, use larger driving forces; (c) to make the engine run more efficiently, use a higher compression ratio.

In contrast, *condensation* may be difficult, time consuming, and very valuable. The objective is to determine whether the particular experimental situation is representative of a class of similar situations. If so, then the measured data might apply to the class: solutions to a class of problems are condensed into the solution of a particular problem. A common way of achieving this kind of generalization is via scaling laws. This approach goes by different names in different disciplines; it is sometimes called universality, sometimes dimensional analysis, sometimes corresponding states. Under any name, the basic idea is to reduce the data to a dimensionless form that is expected to apply to a class of substances or situations. For example, aerodynamic studies of small, model aircraft in wind tunnels can be scaled to predict the aerodynamical behavior of larger, real aircraft.

As another example, we use measurements of the second virial coefficient of gases. Consider again the plot of compressibility factors for steam at 300°C, which appears in Figure 6.11. Since $Z \neq 1$, this steam is not ideal; however, the curve is nearly linear over the range of densities plotted. Therefore we can reliably represent the data in Figure 6.11 by a straight line,

$$Z = 1 + B\rho \tag{6.8}$$

where ρ is the molar density. The quantity B is called the second virial coefficient; it can be measured, has dimensions of reciprocal molar density, and depends only on temperature.

The result in (6.8) implies that there are conditions over which a gas is not ideal ($Z \neq 1$), yet Z remains a relatively simple function of state: it is linear in density along isotherms. (We caution that this behavior is

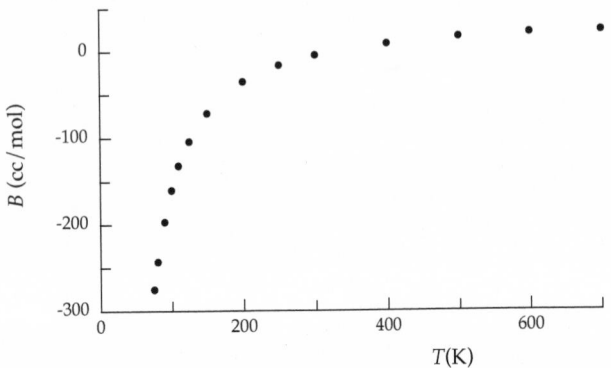

FIGURE 6.12   Second virial coefficients B for nitrogen. Data from Dymond and Smith [4].

restricted to moderately low densities.) To exploit this observation, we turn from steam to a simpler substance. Assume we have conducted PvT experiments on nitrogen, obtaining B as a function of temperature T. The data appear in Figure 6.12. To linearize this data, we follow the procedure in §3.1.1, which leads us to find that B vs $1/T^{3/2}$ is nearly linear, as in Figure 6.13. The correlation in Figure 6.13 should be useful for interpolating among the measured data for nitrogen. The issue now is whether we can do more, without performing more measurements.

To generalize the result in Figure 6.13, we recognize that nitrogen is a small, nonpolar, diatomic molecule. So we might expect that other

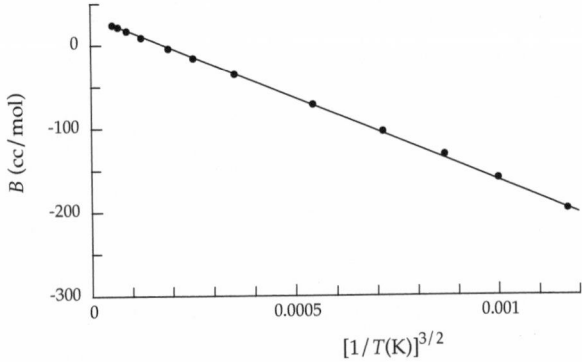

FIGURE 6.13   Second virial coefficients for nitrogen, replotted from Figure 6.12. Line is a least squares fit using equally weighted points.

gases, composed of molecules similar to nitrogen, might have similar values for B, if we can scale the data appropriately. The scaling should leave us with dimensionless forms for B and T.

One possible scaling can be implemented by introducing critical properties, that is, values of properties at the gas-liquid critical point. Here we need the critical temperature T_c and the critical pressure P_c. Then we can form a dimensionless temperature T/T_c and a dimensionless second virial coefficient BP_c/RT_c, where R is the gas constant. In this way we scale the data for nitrogen, obtaining the line in Figure 6.14.

The hypothesis is that the line in Figure 6.14 applies, not to just nitrogen, but to other gases that are also composed of small, nonpolar, nearly diatomic molecules. To test this, we might measure or find second virial coefficients for other gases, scale them to dimensionless form, and add those values to our plot for nitrogen. Such points are shown in Figure 6.14 using data for oxygen, methane, and ethane. The agreement appears good except at the lowest temperatures. From measurements on one substance (nitrogen) we have developed a correlation that estimates property values of other substances. Reliable experiments are expensive and time consuming, so this kind of generalized condensation makes economical use of resources. Furthermore, it also economizes our understanding, for we have quantified the behavior, not of just one substance, but of a class of substances.

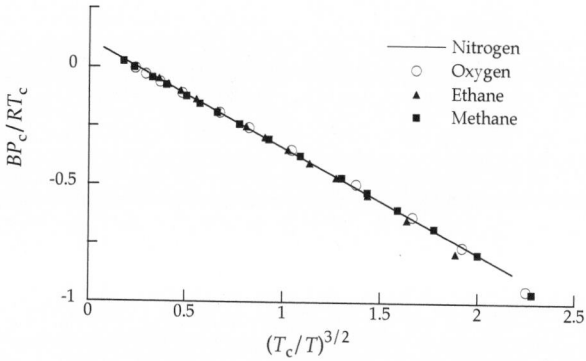

Figure 6.14 Dimensionless second virial coefficients for nitrogen (line taken from Figure 6.13) compared with those for oxygen, methane, and ethane. All data from Dymond and Smith [4].

Literature Cited

[1] K. R. Popper, *The Logic of Scientific Discovery*, Hutchinson and Co., London, 1959.

[2] W. G. Mallard and P. J. Linstrom, eds., *Chemistry Webbook, NIST Standard Reference Database Number 69*, February 2000, National Institute of Standards and Technology, Gaithersburg, MD. Website at http://webbook.nist.gov/chemistry/.

[3] G. Polya, *Mathematics and Plausible Reasoning*, Vol. 1, Induction and Analogy in Mathematics, Princeton University Press, Princeton, NJ, 1954, p. 30.

[4] J. H. Dymond and E. B. Smith, *The Virial Coefficients of Pure Gases and Mixtures*, Clarendon, Oxford, 1980.

Exercises

6.1 From a certain experiment, measurements of y vs x over the range $1 \leqslant x \leqslant 5$, can be correlated by

$$y = a_2 x^2 + a_1 x + a_0$$

where $a_0 = 9.32$, $a_1 = 3.41$, and $a_2 = -0.59$.

(a) Determine the range on x that captures the extremum in y with a tolerance of $\pm 1\%$.

(b) Decompose the curve $y(x)$ into two monotonic curves, $y_1(x)$ and $y_2(x)$, each having the same concavity and whose algebraic sum reproduces $y(x)$, including its extremum.

(c) Cross plot your curves as y_2 vs y_1, locate the point of the extremum, and use your cross-plot to explain whether y_1 or y_2 dominates $y(x)$ on each side of the extremum.

6.2 Assume each of the following conditionals is true. For each, write the converse, the inverse, and the contrapositive forms. Then state whether each of your forms is true or false or whether the given conditional is insufficient to determine the truth value.

(a) If $x = 2$, then $x^2 = 4$.

(b) If $P \geqslant 10$ bar, the tank will rupture.

(c) If $T_2 \geqslant T_1$, then heat flows from object 2 to object 1.

(d) If the data are correct, then the theory is true.

6.3 For linearizing the data for second virial coefficients in Figure 6.12, why don't we use a log-log plot, instead of Figure 6.13?

6.4 In a certain experiment it was found that x and y are correlated by

$$\ln y = 0.916 - 1.50 \ln x$$

(a) Determine the range on x that gives $y = 2$ with a tolerance of $\pm 2\%$.

(b) For a range of $1.95 \leqslant x \leqslant 2.05$, find the corresponding tolerance ($\pm\%$) on y.

6.5 Consider each of the following conditionals and decide whether it is true. Then for each of these, write the converse, the inverse, and the contrapositive forms and state whether each of your forms is true, false, or whether the truth value is undecidable.

(a) If the fire is class A, use water to extinguish it.

(b) If the theory disagrees with the data, the theory is wrong.

(c) If we see a white bear, we are above the arctic circle.

(d) If $r > 0.9$, then a straight line will correlate the data.

(e) If the theory agrees with the data, the data are wrong.

(f) If data from laboratory A agree with data from laboratory B, then the data are correct.

6.6 From a certain experiment, measurements of y vs x over the range $10 \leqslant x \leqslant 24$, can be correlated by

$$y = a_2 x^2 + a_1 x + a_0$$

where $a_0 = 36.85$, $a_1 = -1.21$, and $a_2 = 0.32$.

(a) Determine the range on x that captures the extremum in y with a tolerance of $\pm 3\%$.

(b) Decompose the curve $y(x)$ into two monotonic curves, $y_1(x)$ and $y_2(x)$, each having the same concavity and whose algebraic sum reproduces $y(x)$, including its extremum.

(c) Cross plot your curves as y_2 vs y_1, locate the point of the extremum, and use your cross-plot to explain whether y_1 or y_2 dominates $y(x)$ on each side of the extremum.

6.7 Find two monotonic curves whose algebraic sum reproduces the maximum that appears in Exhibit F.

6.8 The normal boiling points (T_b) and critical temperatures (T_c) of the first five straight-chain alkanes are as follows:

Alkane	Formula	T_b(K)	T_c(K)
Methane	CH_4	111.7	190.6
Ethane	C_2H_6	184.5	305.4
n-Propane	C_3H_8	231.1	369.8
n-Butane	C_4H_{10}	272.7	425.2
n-Pentane	C_5H_{12}	309.2	469.6

Use these data to develop a correlation that relates boiling point to carbon number. Use your correlation to estimate the normal boiling point for n-hexane (C_6H_{14}), which has $T_c = 507.4$ K.

6.9 The following statement is true: If the sample is an ideal gas, then $Z = 1$. Write the converse, inverse, and contrapositive forms of this conditional, and state whether each is true or false. Defend your answers, taking into account Eq. (6.8) and Figure 6.11.

6.10 Values of the second virial coefficient for nitrogen at high temperatures are as follows (from Figure 6.12). Determine the limiting value for B at extremely high temperatures.

T(K)	300	400	500	600	700
B(cc/mol)	−4.2	9.0	16.9	21.3	24.0

6.11 Consider this sequence of integers: 0, 1, 1, 2, 3, 5, 8, 13, . . .

(a) Determine the value for the next integer in the sequence.

(b) Is there any way you can prove that your answer in (a) is correct? If not, how can your answer be confirmed?

(c) Is there a distinction between testing your answer in (a) and confirming your answer? If so, articulate the distinction as clearly as possible.

6.12 It is proposed to determine the second virial coefficient B for a certain pure gas at a specified temperature T. The experimental procedure will be based on the simple equation of state

$$Z = P/\rho RT = 1 + B\rho$$

where $\rho = N/V$, P is the absolute pressure, and R is the gas constant. The procedure is to place a fixed number of moles N of the gas in the cylinder of a piston-cylinder device, then immerse the piston-cylinder in a heat bath that controls temperature to the desired value T. An accurate pressure gage will be attached to the cylinder; the experiment will be performed at pressures up to about 10 bar.

(a) Would better results be obtained (i) by adjusting the position of the piston to produce a specified volume, then measuring the corresponding pressure, or (ii) by adjusting the piston to produce a specified pressure, then measuring the corresponding volume? Defend your answer.

(b) Once Z vs ρ data have been obtained, a curve fit to the above form might be used to obtain a value for B. What consistency checks does the above form suggest could be applied to the measurements?

(c) For the method in (b), how is the uncertainty in the value computed for B related to the uncertainties in Z and ρ?

(d) Is it mathematically possible for errors in the measured values for P to compensate for errors in the measured values for V so that a fortuitously reliable value would be obtained for B? If so, explain how a proper assignment of uncertainties to the measured P and V data might help reveal such a situation.

(e) In the method in (b), a slope is evaluated to obtain B; however, slopes are notoriously difficult to extract from experimental data. Rearrange the above equation so that B would be obtained as an intercept. For your method, how would uncertainties in B be related to uncertainties in Z and ρ?

A

Literature Cited in Exhibit B

[A] A. Cornaz, B. Hubler, and W. Kündig, *Phys. Rev. Lett.*, 72, 1152 (1994).

[B] M. P. Fitzgerald and T. R. Armstrong, *IEEE Trans. Instru. Meas.*, 44, 494 (1995).

[C] H. Walesch, H. Meyer, H. Piel, and J. Schurr, *IEEE Trans. Instru. Meas.*, 44, 491 (1995).

[D] C. H. Bagley and G. G. Luther, *Phys. Rev. Lett.*, 78, 3047 (1997).

[E] J. Luo, Z.-K. Hu, X.-H. Fu, S.-H. Fan, and M.-X. Tang, *Phys. Rev. D*, 59, 042001 (1998).

[F] M. P. Fitzgerald and T. R. Armstrong, *Meas. Sci. Technol.*, 10, 439 (1999).

[G] U. Kleinevoss, H. Meyer, A. Schumacher, and S. Hartman, *Meas. Sci. Technol.*, 10, 492 (1999).

[H] S. J. Richman, T. J. Quinn, C. C. Speake, and R. S. Davis, *Meas. Sci. Technol.*, 10, 460 (1999).

[I] J. P. Schwarz, D. S. Robertson, T. M. Niebauer, and J. E. Faller, *Meas. Sci. Technol.*, 10, 478 (1999).

[J] F. Nolting, J. Schurr, St. Schlamminger, and W. Kündig, *Meas. Sci. Technol.*, 10, 487 (1999).

[K] J. H. Gundlach and S. M. Merkowitz, *Phys. Rev. Lett.*, 85, 2869 (2000).

B

General Linear Least-Squares Problem

CONSIDER THE SITUATION IN WHICH we have measured values for $y(x)$ at N values of x. A plot of the data show that $y(x)$ is nonlinear; however, we expect that $y(x)$ can be represented by some low-order polynomial,

$$y(x) = a_0 + a_1x + a_2x^2 + \cdots + a_mx^m \qquad (B.1)$$

We choose the polynomial to be of a particular order m, then our problem is to determine values for the $(m + 1)$ coefficients a_i. Note that although the data are nonlinear in x, the unknowns are the coefficients a_i and (B.1) is *linear* in those coefficients. Therefore, we can use a linear least-squares procedure to determine the coefficients. The only constraint on this approach is that the number of measurements N must be at least as large (and preferably larger) than the order of the polynomial: $N > m$.

B.1 Least-Squares Equations for a Quadratic Fit

To keep the notation relatively simple, we use this Appendix to develop the least-squares solution for a quadratic polynomial ($m = 2$),

$$y(x) = a_0 + a_1x + a_2x^2 \qquad (B.2)$$

However, what we do here generalizes to polynomials of any order m.

To begin, we form the deviations $\delta(x_i)$ between measured values $y(x_i)$ and the estimates that will be given by the polynomial (B.2),

$$\delta y_i = y(x_i) - (a_0 + a_1x_i + a_2x_i^2) \qquad (B.3)$$

We then seek the values of $a_0, a_1,$ and a_2 that minimize the sum of the squares of the deviations,

$$\underset{\{a_0, a_1, a_2\}}{\text{Min}} \sum_{i=1}^{N} \left(\frac{\delta y_i}{u_i} \right)^2 \tag{B.4}$$

This is analogous to (3.8). Recall that u_i is the uncertainty in the measured value $y(x_i)$. We minimize the quantity in (B.4) by taking the derivative wrt each coefficient and setting each derivative to zero:

$$\frac{\partial}{\partial a_0} \sum_{i=1}^{N} \left(\frac{\delta y_i}{u_i} \right)^2 = 0 \qquad \Rightarrow \qquad \sum_{i=1}^{N} \frac{\delta y_i}{u_i^2} = 0 \tag{B.5}$$

$$\frac{\partial}{\partial a_1} \sum_{i=1}^{N} \left(\frac{\delta y_i}{u_i} \right)^2 = 0 \qquad \Rightarrow \qquad \sum_{i=1}^{N} \frac{x_i \delta y_i}{u_i^2} = 0 \tag{B.6}$$

$$\frac{\partial}{\partial a_2} \sum_{i=1}^{N} \left(\frac{\delta y_i}{u_i} \right)^2 = 0 \qquad \Rightarrow \qquad \sum_{i=1}^{N} \frac{x_i^2 \delta y_i}{u_i^2} = 0 \tag{B.7}$$

Substituting (B.3) for the deviations appearing on the rhs of (B.5)–(B.7), we obtain three linear equations in the three unknown coefficients, $\{a_0, a_1, a_2\}$. We write those equations in a matrix notation:

$$\begin{pmatrix} a_0 \\ a_1 \\ a_2 \end{pmatrix} \begin{pmatrix} S & S_x & S_{xx} \\ S_x & S_{xx} & S_{xxx} \\ S_{xx} & S_{xxx} & S_{xxxx} \end{pmatrix} = \begin{pmatrix} S_y \\ S_{xy} \\ S_{xxy} \end{pmatrix} \tag{B.8}$$

Here S, S_x, S_y, S_{xx}, and S_{xy} are defined in (3.14)–(3.15). Analogously, the remaining symbols in (B.8) are defined as follows:

$$S_{xxx} = \sum_{i=1}^{N} \frac{x_i^3}{u_i^2}, \quad S_{xxy} = \sum_{i=1}^{N} \frac{x_i^2 y_i}{u_i^2}, \quad S_{xxxx} = \sum_{i=1}^{N} \frac{x_i^4}{u_i^2} \tag{B.9}$$

The set of equations in (B.8) can be solved by any method from linear algebra. For example, Cramer's rule yields

$$a_0 = \frac{\begin{vmatrix} S_y & S_x & S_{xx} \\ S_{xy} & S_{xx} & S_{xxx} \\ S_{xxy} & S_{xxx} & S_{xxxx} \end{vmatrix}}{\Delta} \tag{B.10}$$

$$a_1 = \frac{\begin{vmatrix} S & S_y & S_{xx} \\ S_x & S_{xy} & S_{xxx} \\ S_{xx} & S_{xxy} & S_{xxxx} \end{vmatrix}}{\Delta} \tag{B.11}$$

$$a_2 = \frac{\begin{vmatrix} S & S_x & S_y \\ S_x & S_{xx} & S_{xy} \\ S_{xx} & S_{xxx} & S_{xxy} \end{vmatrix}}{\Delta} \tag{B.12}$$

where Δ is the determinant of the coefficient matrix in (B.8),

$$\Delta = \begin{vmatrix} S & S_x & S_{xx} \\ S_x & S_{xx} & S_{xxx} \\ S_{xx} & S_{xxx} & S_{xxxx} \end{vmatrix} \tag{B.13}$$

B.2 Numerical Example

To illustrate use of the above equations, consider Table B.1, which contains measured values for y at $N = 5$ values of x. We want to fit a quadratic to these data. To keep this example simple, we weight all the data equally; i.e., $u_i = 1$ for all i. First, we form the quantities x^2, x^3, x^4,

TABLE B.1 Measured values for $y(x)$ and the associated calculated quantities used in fitting a quadratic to the data.

x	y	x^2	x^3	x^4	xy	x^2y
0	2.7	0	0	0	0	0
1	4.1	1	1	1	4.1	4.1
2	8.8	4	8	16	17.6	35.2
3	18.4	9	27	81	55.2	165.6
4	32.0	16	64	256	128	512
10	66.0	30	100	354	204.9	716.9

xy, and x^2y for each x-y pair. These are also shown in Table B.1. Then we sum each column in Table B.1, obtaining $S_x = 10$, $S_{xx} = 30$, $S_{xxx} = 100$, $S_{xxxx} = 354$, $S_y = 66$, $S_{xy} = 204.9$, and $S_{xxy} = 716.9$. These appear in the bottom row of Table B.1. Now we substitute the values of these sums into (B.10)–(B13) and find $a_0 = 2.806$, $a_1 = -1.081$, and $a_2 = 2.093$; that is, the fitted quadratic is

$$y(x) = 2.806 - 1.081x + 2.093x^2 \qquad (B.14)$$

If we substitute the values of x from Table B.1 into (B.14), we obtain the estimates for y appearing in Table B.2. The differences between the estimates y_e and the measured values y_m are the deviations δy_i; these are also shown in Table B.2. According to (B.5), every unweighted least-squares fit should give a sum of deviations equal to zero. The sum in Table B.2 is not quite zero because of round-off errors incurred by keeping only four significant figures for the coefficients in the fitted quadratic (B.14). The least-squares procedure assures us that the sum of the squared deviations in Table B.2 is smaller than that given by any other quadratic representation of the data.

Examples of least-squares fits to quadratic polynomials appear in Figures 6.1 and 6.2. Fits to fourth-order polynomials appear in Figures 3.14 and 6.8.

TABLE B.2 Deviations and squared deviations between the measured values $y_m(x)$ from Table B.1 and the estimates $y_e(x)$ from the quadratic in (B.14).

x	y_m	y_e	δy	$(\delta y)^2$
0	2.7	2.806	+0.106	0.011
1	4.1	3.818	−0.282	0.080
2	8.8	9.016	+0.216	0.047
3	18.4	18.400	0	0
4	32.0	31.970	−0.030	0.001
		sums:	0.01	0.139

Index

About This Book

This book was typeset in LaTeX using the Memoir class created by Peter Wilson. The text is set in Aldus, the equations in Palatino, and the chapter titles and section headings are in Optima. All three fonts were designed by the master typographer Hermann Zapf.

The figure on the cover is an example of a Lissajous figure: plane curves formed by the intersection of two sine waves that lie orthogonal to one another,

$$x(t) = a_x \sin(b_x t + c_x)$$
$$y(t) = a_y \sin(b_y t + c_y)$$

When the amplitudes, frequencies, and phases of the two waves differ, the resulting figures are complicated, intermeshing curves (as on the cover). In addition, these special cases occur:

1. When the frequencies and phases of the two waves are the same ($b_x = b_y$ and $c_x = c_y$) but the amplitudes differ, the resulting Lissajous figure is a straight line passing through the origin. If the amplitudes are also the same ($a_x = a_y$), then the straight line lies at $45°$ to the x and y axes.

2. When the frequencies and amplitudes are the same ($b_x = b_y$ and $a_x = a_y$) but the phases differ, the Lissajous figures are ellipses. However, if the phase difference is $90°$ or $270°$, the ellipses collapse to circles.

The figures were first discovered by the American mathematician Nathaniel Bowditch (1773–1838) in 1815, so they are sometimes called Bowditch curves. They were discovered independently by the French mathematician Jules-Antoine Lissajous and studied thoroughly by him in 1857–58.

To create the Lissajous figure appearing on the cover, we dampened the amplitudes:

$$x(t) = a_x \exp[-k_x t] \sin(b_x t + c_x)$$
$$y(t) = a_y \exp[-k_y t] \sin(b_y t + c_y)$$

The resulting motion models that of a decaying, compound pendulum; that is, a pendulum that can move independently in both the x and y directions.